EAI/Springer Innovations in Communication and Computing

Series editor
Imrich Chlamtac, European Alliance for Innovation, Gent, Belgium

Editor's Note

The impact of information technologies is creating a new world yet not fully understood. The extent and speed of economic, life style and social changes already perceived in everyday life is hard to estimate without understanding the technological driving forces behind it. This series presents contributed volumes featuring the latest research and development in the various information engineering technologies that play a key role in this process.

The range of topics, focusing primarily on communications and computing engineering include, but are not limited to, wireless networks; mobile communication; design and learning; gaming; interaction; e-health and pervasive healthcare; energy management; smart grids; internet of things; cognitive radio networks; computation; cloud computing; ubiquitous connectivity, and in mode general smart living, smart cities, Internet of Things and more. The series publishes a combination of expanded papers selected from hosted and sponsored European Alliance for Innovation (EAI) conferences that present cutting edge, global research as well as provide new perspectives on traditional related engineering fields. This content, complemented with open calls for contribution of book titles and individual chapters, together maintain Springer's and EAI's high standards of academic excellence. The audience for the books consists of researchers, industry professionals, advanced level students as well as practitioners in related fields of activity include information and communication specialists, security experts, economists, urban planners, doctors, and in general representatives in all those walks of life affected ad contributing to the information revolution.

About EAI

EAI is a grassroots member organization initiated through cooperation between businesses, public, private and government organizations to address the global challenges of Europe's future competitiveness and link the European Research community with its counterparts around the globe. EAI reaches out to hundreds of thousands of individual subscribers on all continents and collaborates with an institutional member base including Fortune 500 companies, government organizations, and educational institutions, provide a free research and innovation platform.

Throughs its open free membership model EAI promotes a new research and innovation culture based on collaboration, connectivity and recognition of excellence by community.

More information about this series at http://www.springer.com/series/15427

Rui José • Kristof Van Laerhoven
Helena Rodrigues
Editors

3rd EAI International Conference on IoT in Urban Space

Springer

Editors
Rui José
Algoritmi Research Centre
University of Minho
Braga, Portugal

Helena Rodrigues
Braga, Portugal

Kristof Van Laerhoven
Siegen, Nordrhein-Westfalen, Germany

ISSN 2522-8595 ISSN 2522-8609 (electronic)
EAI/Springer Innovations in Communication and Computing
ISBN 978-3-030-28924-9 ISBN 978-3-030-28925-6 (eBook)
https://doi.org/10.1007/978-3-030-28925-6

This Springer imprint is published by the registered company Springer Nature Switzerland AG.
The registered company address is: Gewerbestrasse 11, 6330 Cham, Switzerland

Preface

We are delighted to introduce the proceedings of the third edition of the 2018 European Alliance for Innovation (EAI) International Conference on IoT in Urban Space (Urb-IoT), co-located with the Smart City 360° Summit 2018, which took place in Guimarães, Portugal. This conference has brought together researchers, developers, and practitioners around the world who are exploring the urban space and its dynamics within the scope of the Internet of Things (IoT) and the new science of cities.

The technical program of Urb-IoT 2018 consisted of 15 full papers, which have been distributed by four parts: Part I—IoT for Urban Analytics; Part II—Design Challenges in Urban IoT; Part III—Methods and Technologies for IoT; and Part IV—Urban Mobility and Connected Vehicles.

Coordination with the steering chairs, Prof. Imrich Chlamtac and Prof. Fahim Kawsar, as well as with the General Chair of Smart City 360° Summit, Prof. Henrique Santos, was essential for the success of the conference. We sincerely appreciate their constant support and guidance. It was also a great pleasure to work with such an excellent organizing committee team for their hard work in organizing and supporting the conference: in particular, the Technical Program Committee, led by our TPC Co-Chairs, Prof. Rui José and Prof. Kristof Van Laerhoven, and all the additional reviewers, who have completed the peer review process of technical papers and made a high-quality technical program. We are also grateful to Conference Managers Kristina Lappyova, Marek Kaleta, and Andrea Piekova for their support.

Finally, a special mention to all the authors who submitted their work to the Urb-IoT 2018 conference. We strongly believe that Urb-IoT conference provides a good forum for all researchers, developers, and practitioners to discuss all science and technology aspects that are relevant to smart cities. We are now even more confident that future Urb-IoT conferences will be as successful and stimulating as suggested by the contributions presented in this volume.

Braga, Portugal — Rui José
Siegen, Germany — Kristof Van Laerhoven
Braga, Portugal — Helena Rodrigues

Conference Organization

Steering Committee	
Imrich Chlamtac	European Alliance for Innovation, Gent, Belgium
Fahim Kawsar	Bell Labs, Belgium
Organizing Committee	
General Chair	
Helena Rodrigues	University of Minho, Portugal
TPC Co-Chairs	
Rui José	University of Minho, Portugal
Kristof Van Laerhoven	University of Siegen, Germany
Local Chair	
Maria João Nicolau	University of Minho, Portugal
Publicity and Social Media Co-Chairs	
Fábio Silva	Polytechnic Institute of Porto, Portugal
Vítor Santos	New University of Lisbon, Portugal
Publications Chair	
Filipe Meneses	University of Minho, Portugal
Web Chair	
César Analide	University of Minho, Portugal
Posters and Demos Chair	
Ana Aguiar	University of Porto, Portugal
Conference Manager	
Kristina Lappyova	EAI
Marek Kaleta	EAI
Technical Program Committee	
Adriano Moreira	University of Minho
Afra Mashhadi	University of Washington
Akihito Sudo	Shizuoka University
Fahim Kawsar	Nokia Bell Labs
Florian Michahelles	Siemens Corporation

Graça Carvalho	UCL
Kota Tsubouchi	Yahoo! Japan Research
Lawrence O'Gorman	Nokia Bell Labs
Matthias Budde	Karlsruhe Institute of Technology
Naonori Ueda	NTT Communication Science Laboratories
Nuno Lopes	United Nations University
Philipp M. Scholl	University of Freiburg
Raffaele Giaffreda	FBK CREATE-NET, Italy
Rossi Kamal	Shanto Mariam University of Creative Technology
Sarah Gallacher	Arup
Stefan Van Der Spek	Delft University of Technology
Sumi Helal	Lancaster University
Susanna Pirttikangas	University of Oulu
Takuro Yonezawa	Keio University
Till Riedel	Karlsruhe Institute of Technology, Germany
Ulf Blanke	ETH Zürich
Yoshihide Sekimoto	University of Tokyo
Yoshito Tobe	Aoyama Gakuin University
Additional Reviewers	
Ana Aguiar	University of Porto
Alexander Hoelzemann	University of Siegen
Andreas Wiebe	University of Siegen
César Analide	University of Minho
Christopher Bull	Lancaster University
Fábio Silva	Polytechnic Institute of Porto
Florian Wolling	University of Siegen
Helena Rodrigues	University of Minho
Maria João Nicolau	University of Minho
Jochen Kempfle	University of Siegen
Mahsa Honary	Lancaster University
Vitor Santos	New University of Lisbon

Contents

About the Editors

Rui José is Assistant Professor at the Department of Information Systems, School of Engineering, University of Minho. He holds a PhD in Computer Science, from Lancaster University, UK (2001). He is also a full member of Centro Algoritmi, where he leads the Mobile and Ubiquitous Systems group, which is part of the research line in Computer Communications and Pervasive Media. His research interests include ubiquitous systems and smart cities, more specifically the design and development of urban information services that are situated within particular physical and social settings. He has over 130 publications in books, book chapters, journals, and proceedings of international and national conferences. He has contributed to top journals and conferences, such as Ubicomp Conferences, IEEE Computer, IEEE Computer Graphics and Applications, IEEE Pervasive Computing, and Sensors. Rui José has been extensively involved in collaborative research, more specifically 12 research projects with competitive funding and 10 industry contracts. He has been PI or co-PI on 8 projects, including projects funded by international programs, such as FP7, FET-OPEN, and Portugal/CMU.

Kristof Van Laerhoven is full Professor of Ubiquitous Computing at the University of Siegen (Germany) and studies systems that combine wirelessly connected and wearable sensors with embedded machine learning techniques. Previously, he was Professor of Embedded Systems at the University of Freiburg and has led the Embedded Sensing Systems Research Group at the Technical University of Darmstadt. He obtained his PhD at the University of Lancaster (UK) and studied Computer Science at the University of Brussels (VUB) in Belgium. He has chaired the premier International Conference for Wearable Computing, ISWC, is active as Associate Editor for the *Ubiquitous Computing IMWUT* journal, and heads the wearables department of the *IEEE Pervasive Computing* magazine. More info on: http://ubicomp.eti.uni-siegen.de

Helena Rodrigues is an Assistant Professor at the University of Minho in Portugal, where she is a member of the Urban Informatics research group of the Algoritmi Research Center. She did her undergraduate studies and received her MSC degree in Computer Science at Universidade do Minho, Portugal. She received her PhD in

Computer Science from the University of Kent, UK, in 1999. Her current research interests include methodologies and cloud-based platforms for the development of applications for mobile and IoT-based systems and applications. Helena Rodrigues has participated in several research projects in the field of mobile and ubiquitous computing, smart cities, and industry 4.0. She has published her research results in different conferences and journals in the area. She has also participated in several research meetings and conferences. She has recently served as General Chair of Urb-IoT 2018—3rd EAI International Conference on IoT in Urban Space and has served as reviewer for a number of international journals and conferences. She is supervising PhD and master's students and has participated in PhD and master's examinations.

Part I
IoT for Urban Analytics

CityFlow: Supporting Spatial-Temporal Edge Computing for Urban Machine Learning Applications

Makoto Kawano, Takuro Yonezawa, Tomoki Tanimura, Nam Ky Giang, Matthew Broadbent, Rodger Lea, and Jin Nakazawa

1 Introduction

How can we develop machine learning applications to make the city smarter? In order to realize *smart cities*, there has been extensive research into using machine learning (ML) techniques. For example, citizens' behavior prediction by using GPS trajectories of their smart devices or taxis [18], and road damage detection from video recorders mounted on automobiles [9]. In both cases, city data is gathered into centralized cloud servers for analysis. This centralized approach is well suited to ML algorithms, but unfortunately does not reflect the reality of Smart Cities. Typically, cities have a multitude of sensors and "things" (e.g., automobiles, mobile devices, and robots) that are distributed throughout the city and communicate with each other via the Internet (e.g., Wi-Fi, LTE, 3G, or Ethernet). This results in city data exhibiting a wide variation in time and space both in short term, i.e., daily cycles, and in longer term cycles as the city, its infrastructure, and its citizens evolve. In short, the city is a spatial-temporal distributed edge environment (STDEE) and requires we adapt our ML techniques to this environment. However, developing and operating ML applications in an STDEE is not straightforward. This is primarily

M. Kawano (✉) · T. Yonezawa · T. Tanimura · J. Nakazawa
Keio University, Fujisawa, Kanagawa, Japan
e-mail: makora@ht.sfc.keio.ac.jp; takuro@ht.sfc.keio.ac.jp; tanimu@ht.sfc.keio.ac.jp; jin@ht.sfc.keio.ac.jp

N. K. Giang
University of British Columbia, Vancouver, BC, Canada
e-mail: kyng@ece.ubc.ca

M. Broadbent · R. Lea
Lancaster University, Bailrigg, Lancaster, UK
e-mail: m.broadbent@lancaster.ac.uk; r.lea1@lancaster.ac.uk

R. José et al. (eds.), *3rd EAI International Conference on IoT in Urban Space*, EAI/Springer Innovations in Communication and Computing,
https://doi.org/10.1007/978-3-030-28925-6_1

because the development process, which is already complicated because of the need for multiple trials to aid learning, needs to also accommodate the distributed nature of the STDEE. Obviously this leads to significant development costs, both in terms of time and resources, and is by nature inherently complex.

To ease both the cost and complexity, an integrated development environment is desired for ML application development suitable for STDEEs as found in smart cities. To solve this problem, we have designed and implemented *CityFlow*, which supports the development of ML applications in the STDEE city. CityFlow makes it possible to easily describe the flow of data which comes from devices in the city, and to deploy trained ML models to a variety of devices distributed throughout the city. By making these processes easy, we can conduct data preprocessing and preservation quickly and repeat *proof of concept* (PoC) that verify the performance of the ML model. CityFlow is built using Distributed Node-RED (DNR) [4, 5]. Node-RED is a visual tool for the IoT, interconnecting hardware devices, Web service APIs, and online services quickly. DNR is an extension of Node-RED that supports an edge (or fog) computing environment.

This paper reports two case studies we conducted to evaluate the usefulness of CityFlow. The first study develops a road damage detection application which identifies damaged road and road markings from video images gathered by city garbage trucks. This application utilizes an object detection method via deep neural networks and also protects privacy in visualization. The second study develops an automatic labeling application that allows city officials to report problems which they find in the city, such as the illegal dumping, uncollected garbage, graffiti, etc. Images attached to the reports are classified automatically into the type of problem making reporting easier and less error prone for city officials. The contributions of this paper are summarized as follows:

- We identify key issues in the development and operation of ML applications in spatial-temporal distributed edge environments, such as smart cities.
- We design and implement CityFlow, the integrated development environment to address those issues.
- We validate our approach by reporting on two real world case studies we have designed and deployed to aid city staff.

2 Urban Machine Learning Applications

2.1 Spatial-Temporal Distributed Edge Environments

With the developments in IoT technologies, an enormous amount of data from sensors in the city can be obtained. This covers traditional city infrastructure such as water, electricity, and lighting but also increasingly comes from smart devices such as smart transportation, smart homes, and offices and perhaps most importantly citizen data from their personal devices. At the same time, an increasing number of small edge devices such as embedded computers being deployed: such as NVIDIA Jetson TX2 (embedded computers), Google Embedded TPU, and Xilinx Spartan6 (FPGA).

As the capabilities of these edge devices increase, regardless of their low price, it is possible to use them to support distributed, ML applications. For example, they are already used in domains such as agriculture.[1] Simultaneously, some work has begun to explore deep learning on edge computers [6] for mobile applications. Finally, an important trend is towards treating city information from web service[2] and also the statistics of the city from open data thanks to the spread of IT technology, such as cloud computing technology, which treats the data sources as virtual sensors.

While these trends are leading to significantly more city data being gathered, it is also clear that the spatial-temporal nature of that data is changing. Significant amounts of data are spatial in nature, e.g., where sensors are mounted on garbage trucks, the data is gathered from a variety of locations as the truck travels through the city. Equally, data is gathered during the working day, but not at night—hence it exhibits a high degree of temporal variance. This variety is common to many sources of data, for example data from citizens' mobile devices.Consequently, cities where the IoT and the edge computing technologies are spread throughout exhibit key characteristics of STDEEs.

2.2 *Life Cycle of Machine Learning Applications*

In general, the phases (or life cycle) which are conducted in ML application development are shown in Fig. 1. In the first phase, in order to tackle a particular city problem, data analysis such as statistical estimation is carried out for understanding

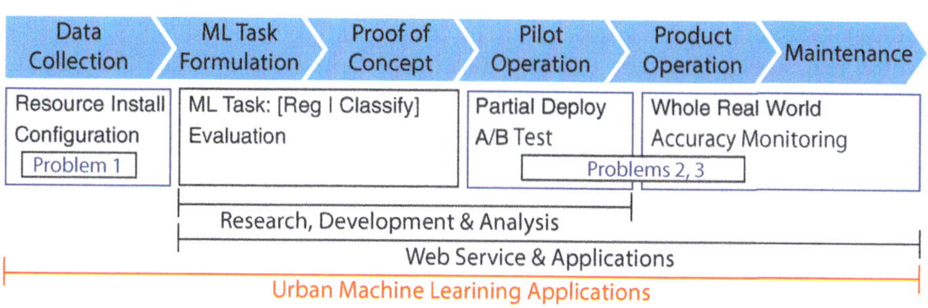

Fig. 1 General life cycle of machine learning applications. There are several types of application to serve their purpose: research, web services, urban application

[1]https://cloud.google.com/blog/products/gcp/how-a-japanese-cucumber-farmer-is-using-deep-learning-and-tensorflow.

[2]http://soramame.taiki.go.jp/.

the gist of the problem and formulate it as an ML task. Most of the task formulation becomes either regression tasks or classification tasks. Simultaneously, the ML approach would be determined by whether the data is labeled (or annotated) fully, partly, or not at all: supervised learning, semi-supervised learning, unsupervised learning.

In the second phase, to perform the formulated ML task, we design and train the ML model using a given dataset. Then, we experiment to evaluate the model performance as regards its accuracy. In this paper, we refer to this process as the Proof of Concept (PoC). If the model performance is poor, it may be necessary to start over by re-formulating the problem.

As a result of the PoC, if the ML model is able to perform adequately, then we implement and install this ML model into the edge environment and verify whether the model is effective in the real world (we refer to this as the pilot operation). During this phase we adjust for accuracy, identifying factors such as data quality or quantity, which can be improved or factors such as the capacity of the model which may result in re-design of the model.

Finally, we shift to the product operation phase during which, in order to improve the model, we periodically re-build the dataset with additional new data which continues through to the maintenance phase.

In summary, in the development and operation of the ML application, we repeat the exploratory phases {task formulation, PoC, pilot deployment, production operation, and maintenance}.

2.3 *The Problems for Machine Learning Applications in the City*

While urban ML applications which are offered to solve the urban problems (e.g., parking availability monitoring [2]) are proposed, there is some difficulty to deploy to the real world. The data used by those applications are collected by sensors which are installed temporarily, with limited coverage, and in a precise controlled manner. Furthermore, they have to select the sensor devices which satisfy their application requirements. In this paper, we call these technical concerns *spatial-temporal device-data dependency*, which occurs in the phase surrounded by blue line in Fig. 1.

Problem 1: Urban Data Collection and Preprocessing In statistical machine learning, it is generally assumed that the data have already been collected, formatted, and normalized to use for the training dataset—often using offline processing and assuming homogenous datasets. In contrast, cities exhibit highly heterogeneous datasets, which change over time as the city changes and where data is streamed in real time. This results in increased processing and storage and requires a dynamic learning model that retrains as data changes. While there exist a number of algorithms which are able to handle streaming data, such as CQL [1], Apache Spark [17], and Storm [14] they often assume cloud based servers with load balancing and scaling, and work on homogenous datasets. Which, as discussed, is not often the case with real world city datasets.

In addition, we usually conduct some pre-(post-)processing of the data. For example, for privacy protection, since the data from edge devices is often bound to the real world, there is a risk of privacy invasion. For example, camera data from city sensors or car data captured in real time has significant privacy issues. However, the diversity of devices makes it hard to cope with the data processing required for dealing with this privacy invasion.

Problem 2: The Machine Learning Model Execution Environment After the PoC phase, it is necessary to distribute the trained ML model to all (or a subset) of the edge devices in the city. In typical ML applications, high-performance cloud computers are utilized. This is often necessary because the ML model has a large number of parameters requiring significant memory and high-performance CPUs. However, it is often too expensive for cities to own and manage such servers due to budget constraints. In contrast, we can use edge devices in a STDEE city that typically have significantly lower specifications than cloud based servers. However, while we can use traditional approaches on edge devices with reasonable computational resources, the latest approaches which have good performance, such as deep neural networks, are not always appropriate due to their resource needs. In order to benefit from their performance, we need to divide the ML model into subsets of small ML model or processes that can be distributed across a set of edge devices [7, 10] or compress it to load it on the edge device memory [6].

Additionally, since the edge devices are highly heterogeneous, it is not just a simple matter of distributing partial modes to a set of edge devices, rather the constraints and context of each device needs to be considered as part of any distribution algorithm.

Problem 3: Covariate Shift of the City Data In statistical machine learning, typically it is required that there is no difference between the marginal probability distribution of the training dataset and that of the test dataset. Namely, assuming the distribution of training dataset is p and that of test dataset is p', $p(\mathbf{x}) = p'(\mathbf{x})$ is required. In an STDEE based city, the network status and the context of the location where the edge devices are installed, changes. Therefore, the marginal distribution becomes different, although the relationship between the data $\mathbf{x}$ and the desired output y that is posterior $p(y|\mathbf{x})$ is consistent. This phenomenon is known as *covariate shift* [12, 13]: $p(\mathbf{x}) \neq p'(\mathbf{x})$, $p(y|\mathbf{x}) = p'(y|\mathbf{x})$. When considering smart cities, we use the hypothesis that the covariate shift exists in the city, as the distribution of the data from the city often varies significantly in spatial and in temporal domains. We call the covariate shift caused in the city as Spatial-Temporal covariate shift.

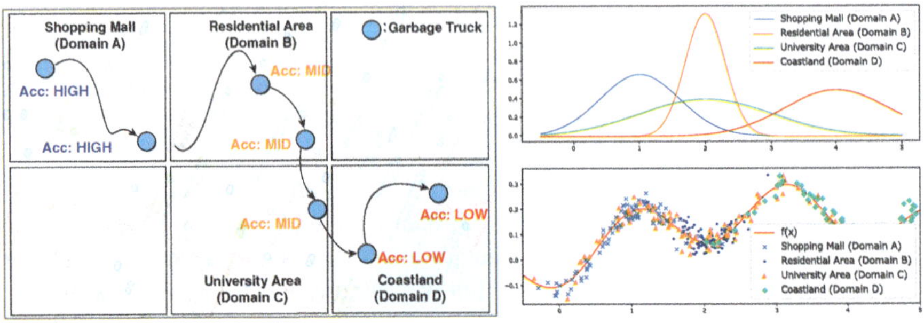

Fig. 2 Influence on accuracy due to a spatial-temporal covariate shift. For models taught from data around commercial facilities, prediction accuracy is high in similar commercial facilities. The prediction accuracy decreases as the spatial features differ

The covariate shift of the city data is illustrated in Fig. 2. The prediction accuracy of the model is high in the shopping street, because the model is trained with the data collected in the shopping street. In contrast, in other areas away from the shopping center, the more the data distribution differs from that of the shopping street, the lower the prediction accuracy is owing to the ST covariate shift. For instance, supposing car detection based on video image analysis is required, a model which is trained with the video of shopping streets can detect cars with high accuracy in similar streets. By contrast, at the sea coast, the model struggles to precisely detect cars because the background is sea water instead of buildings. Similarly, if the model is trained with the videos taken in daytime, it is difficult for it to detect cars during the night, even though the location is the same. Therefore, it is important to consider these ST covariate shifts when we develop the ML applications for the smart city.

3 CityFlow

To remedy the spatial-temporal device-data dependency, we propose CityFlow, which is the system combining *Distributed Node-RED* and *Sensor over XMPP*. The overview is depicted in Fig. 3.

3.1 Design Policy

In order to address the issues described previously, we construct an integrated development environment, called *CityFlow*. This can be used to flexibly realize the life cycle of ML application development in the city. While typical ML applications utilize a centralized processing system, it is difficult to effectively handle the large variety of data sent from the enormous number of edge devices

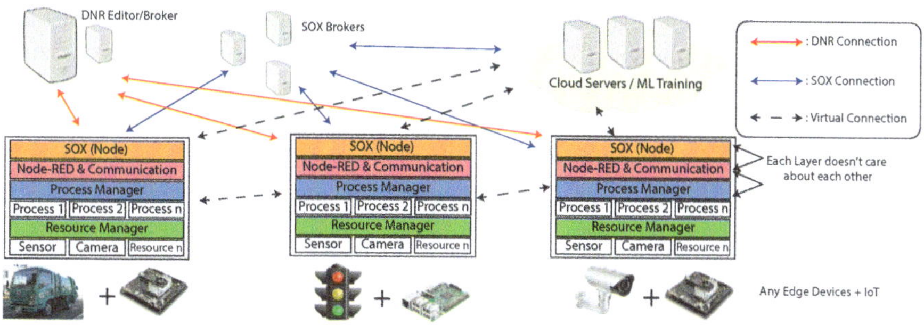

Fig. 3 Overview of CityFlow

in the city. In contrast, the ML applications, developed within CityFlow, adopt a distributed processing approach, better matched to the spatial-temporal distributed edge environment within a city.

Edge Devices and Preprocessing Virtualization While the typical ML lifecycle assumes that the data has been already collected into databases, the same does not apply in a city environment. To tackle this, CityFlow collects the data by controlling the edge devices and facilitating the communication between those devices. When collecting the data, it is important to consider the difference between the specification of edge devices and their installed environment. For instance, the data resource could be a virtual sensor such as a web service API or a physical sensors such as an accelerometer. Although the preprocessing should be performed at the sensor where the data is collected, the format of data often differs between platforms, creating additional complexity.

In order to deal with this, CityFlow abstracts away the different edge devices and different data formats to ensure interoperability. Thus, developers are able to concentrate on developing ML models without concern for the underlying technologies and protocols. Furthermore, this abstraction realizes simplified sharing and reusing of these resources among additional developers or applications.

Resource Management and Efficiency Improvement The task formulation phase and PoC phase in the ML lifecycle is essential to start developing a ML application. In general, it is necessary to select the required data from databases to shift the phase from the task formulation to PoC. However, the presence of data which is not used results in inefficiencies. Therefore, it is efficient for networks and storage mediums to filter the necessary data before storing it into databases. There are several ways to filter data, including:

- **Explicit selection.** Specify the unique device ID to obtain data from it.
- **Implicit selection.** Specify by spatial and temporal range, such as geolocation or time span, etc., and collect only data from devices in that range.

The capability of an edge device, such as CPU and memory, is often lower than that used in cloud computing. Consequently, they cannot perform data preprocessing or model prediction in situations where an edge device receives an enormous amount of data. To effectively handle these situations, the device sends the data to multiple nearby edge devices to balance the load.

Domain Adaptation for Spatial-Temporal Covariate Shift It is necessary to handle the ST covariate shift to develop ML applications in the city scenario described above. However, there are two challenges:

Firstly, since the statistical machine learning including deep neural networks cannot perform extrapolation, we need to take the approach that treats the extrapolated data: ST covariate shift. For instance, we can adopt *domain adaptation learning* which is proposed by Ganin et al. [3]. They assume that a covariate shift occurred between training data (source domain) and test data (target domain). With this assumption, they provide three types of models: the *feature extractor*, which outputs the feature from the common probability distribution wherever the input from the source domain or the target domain, the *solving model* which outputs values for the task (regression or classification), and the *discrimination model* which determines whether the input comes from the source or the target. Although adopting these models is one of the efficient solutions for domain adaptation between two domains, the number of domains in the city might be more than two; it is difficult for the city to adopt these models without change. Therefore, to solve the task, CityFlow is required to distribute the models to edge devices in each domain.

The second challenge is the detection of ST covariate shift, which is an estimation of the boundary between the domains as shown in Fig. 2. In typical ML applications, the meta-information, such as the device IDs and locations, is attached to the data because of being centralized; this is problematic for treating ST covariate shift. Accordingly, CityFlow is required to provide a function that is capable of monitoring the performance of ST covariate shifting and estimating when the boundary occurs.

3.2 Implementation

Distributed Node-RED Node-RED is a dataflow-based visual programming tool and language for IoT applications. Dataflow is a natural programming model represented as a directed graph of processing nodes, each of which representing a resource. The applications are developed by dragging and dropping processing nodes onto a canvas, and "wiring" the nodes together. The wires represent communication paths between nodes. The resulting application is referred to as a "flow", which can be deployed to a single device. However, Node-RED has no support for distributed edge environments.

Distributed Node-RED (DNR) is an extension of Node-RED for distributed environments [4, 5]. There are three notable extensions, which are described below.

The first extension is the notion of a *device* within the dataflow language. This notion enables individual hardware devices to be uniquely identified in a distributed environment. The device also carries a set of characteristics. This allows the user to determine which device(s) a node should be deployed to and executed on; for example, mobile devices, embedded computers, and cloud servers.

The second extension is the notion of *remote wires*. *Remote wires* make it possible to support inter-device communication so that the nodes running on distributed separate devices can send the data. These wires are implemented by using a publish/subscribe communication mechanism. For instance, assume that node A sends the data to node B. Then node A publishes the data to the communication broker so that node A does not need to know where node B is. Similarly, node B subscribes to the broker so that node B can receive the data without needing to know its source node. Additionally, using *remote wires* logically separates the process so that highly computational processes can be conducted on the multiple edge devices without using cloud computing.

The last extension is the *constraint* primitive, used as a broader abstraction that specifies how a node is deployed and run in a distributed setting. Accordingly, every node in a dataflow is attached with a *constraint*, property that defines how the deployment is conducted. This *constraint* indicates requirements on device identification, computing resources such as CPU and memory, and physical location. For example, by using the *constraint*, we could restrict a sensor processing flow so that it only runs on a device mounted on a moving vehicle, located within a certain geographic area.

By using DNR for CityFlow, developers can concentrate on developing ML applications without knowing which devices and data resources are used. For more details or other extensions of DNR, see [4].

Sensor-Over-XMPP Node CityFlow offers a node implemented in DNR that handles Sensor-Over-XMPP (SOX). SOX is the specification of SOXFire [16] which is a universal sensor data exchange system. This utilizes the Internet protocol of the open XML format (XMPP), typically used for chat communications, to represent the meta-information. Accordingly, by using SOX nodes, the data can be treated uniformly from the physical sensors and the virtual sensors. This feature is useful for the data used in ML techniques in the city. Furthermore, since the ML model is distributed to the edge devices all over a city to conduct prediction locally, SOX can be used.

4 Case Study

4.1 Road Damage Detection

In this paper, we use a road damage detection application as a first case study, with a focus on road line markings. Although road damage is a common problem, in

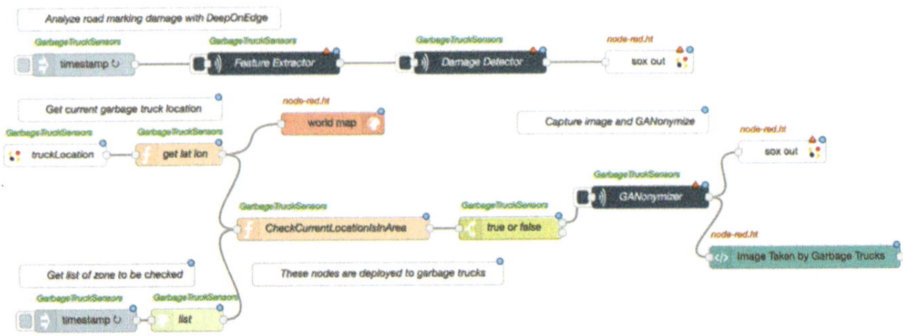

Fig. 4 Road damage detection application developed by CityFlow

many cities, inspection is still conducted by sight. This manual visual inspection is resource intensive and expensive. Therefore, we have to explore ways to inspect the city infrastructures, especially roads, at low cost. Recent work by Kawano et al. [9] proposed a method for road inspection using recorded images from cameras mounted on garbage trucks. However their approach relied on central cloud processing which is inefficient and costly in terms of processing and communications and fails to handle local privacy issues (*Problem 1*). Simultaneously, since they adopted an object detection approach [11] which uses a large neural network, it is difficult to work on edge devices (*Problem 2*). Moreover, the cityscapes of images changes in some areas, so that the ST covariate shift might occur between those areas (*Problem 3*). Our application to address these problems, implemented using CityFlow, operates as follows in Fig. 4.

The Flow of Fig. 4 Top In order to deploy the networks to edge devices and cope with ST covariate shift, the neural network is separated into *feature extractor* node and *damage detector* node and deployed to different devices on the truck, respectively. Consequently, we can adapt *domain adaptation* approach [3]. Then, when the damaged road is detected through these networks nodes, the location of it is published to SOXFire via SOX Node (SOX-out in the flow).

The Flow of Fig. 4 Bottom When multiple garbage trucks confirm an area of road damage, one truck is designated to upload a partial video of the area. A SOX-in node receives the location information, and the next node compares the location of the truck with it. If it is true, the driving recorder mounted on the truck sends the videos after anonymizing to the cloud visualization application. Before uploading, information related to the person such as faces or car matriculation plates is removed if the video contains them. Upload and display the videos in a suitable application for confirmation by city staff.

4.2 Automatic Labeling in Participatory Sensing

The second case is the participatory sensing application for city staffs. We assume a system to report when the city staffs discover issues with household garbage such as illegal dumping, uncollected garbage, graffiti, etc. Currently, it is used by city staff and the information from reports is gathered daily in several cities in Japan. However, the system requires city staff members to manually input the information about the issue types (12 classes), the degree of urgency (emergency/normal/nothing), and a free comment. This often leads to inconsistencies.

To address this, we have designed an application that automatically labels and makes the report based on image recognition techniques. Given the automatic labeling application, we need to preserve the model quality to be able to label precisely. However, there is no generic ML model (No-free-lunch theorem), so that the model should be selected to suit the area (*Problem 3*) and work on their mobile device (*Problem 2*). In addition, we want to know the situation of specific areas in some cases. To accomplish this, the system requires the city staff members who are in the specific area to check those surroundings (*Problem 1*).

This application flow also can easily be implemented using CityFlow thanks to GUI that can represent the logic intuitively and bind them to devices. The application automatically classifies the photo images taken from the camera by each city staff member. The automatic labeling of the captured image is conducted by the image recognition model. If the result of automatic labeling is incorrect, the city staff member revises it. Then, the information is published by SOX-out nodes. When the number of reports about the specific area saved to SOXFire exceeds an arbitrary threshold, the system updates the dataset, so that the model performance improves in that area by re-training the model in the cloud servers. It is expected that as more data is gathered, the model performance improves in the specific area.

5 Related Work and Discussion

This paper presents a novel system called CityFlow, which allows us to build urban ML applications easier. Although there are a number of streaming process engines using multiple devices [1, 14, 17], they are unable to filter the data according to application spatial or temporal constraint. In contrast, CityFlow can filter explicitly or implicitly, allowing us to collect data flexibly and efficiently, thus better supporting STDEE. While DeepMon [7] or DeepX [10] make it possible to execute deep neural networks on mobile devices by distributing the networks to multiple processes, they only support neural networks. In contrast CityFlow is agnostic with respect to ML approach and supports other processes such as data preprocessing and filtering by using DNR. This enables CityFlow to support the complete end-to-end urban application development and, by exploiting DNR's distribution and edge processing capabilities, support a complex ML approach which demands significant processing. However, at this stage, CityFlow does not support model training and labeling of the dataset in the STDEE. We utilized the NVIDIA Jetson TX2 in the case study, which is one of the highest specification edge computers that offers a GPU. Although it can perform the fast inference with deep neural networks, it is difficult to use for model training due to the low capacity of their RAM.

In addition, we assume data annotation has already done by hand, since preparing the fully labeled dataset and adopting supervised learning is the best approach with respect to solving the ML task. However, because the cost of annotation has been highlighted as a significant issue for ML, we have to explore ways to make better use of human annotation [8, 15].

6 Conclusion and Future Work

We presented CityFlow, an integrated development environment system for developing urban ML applications which is built using Distributed Node-Red and Sensors over XMPP. We have outlined some of the key issues which arise in developing urban ML applications due to the city being a spatial-temporal distributed edge environment. To show how CityFlow is useful, we conducted two case studies for city infrastructure and service improvement. Our implementations highlighted the flexibility in data capture and processing that CityFlow offers and demonstrated that a dataflow model, that virtualizes ML steps as dataflow nodes, makes developing applications easier. For future work, we are planning to conduct measuring the time to implement the urban ML application by programmers or city officers for the quantitative evaluation and also taking a survey for the qualitative evaluation. Moreover, we plan to evaluate our applications with CityFlow over a longer period of operation to understand better the nature and presence of ST covariate shift in the city. Additionally, we hope to address the issues of distributed model training which we highlighted earlier.

Acknowledgements This work was supported in part by National Institute of Information and Communications Technology and in part by H2020-EUJ-2016 EU-Japan joint research project, BigClouT (Grant Agreement No 723139).

References

1. Arasu, A., Babu, S., Widom, J.: The CQL continuous query language: semantic foundations and query execution. VLDB J **15**(2), 121–142 (2006)
2. Araújo, A., Kalebe, R., Girað, G., Gonçalves, K., Neto, B., et al.: Reliability analysis of an IoT-based smart parking application for smart cities. In: 2017 IEEE International Conference on Big Data (Big Data), pp. 4086–4091. IEEE, Piscataway (2017)

3. Ganin, Y., Ustinova, E., Ajakan, H., Germain, P., Larochelle, H., Laviolette, F., Marchand, M., Lempitsky, V.: Domain-adversarial training of neural networks. J. Mach. Learn. Res. **17**(1), 2096–2030 (2016)
4. Giang, N.K., Lea, R., Leung, V.C.M.: Exogenous coordination for building fog-based cyber physical social computing and networking systems. IEEE Access **6**, 31740–31749 (2018). https://doi.org/10.1109/ACCESS.2018.2844336
5. Giang, N.K., Lea, R., Blackstock, M., Leung, V.C.M.: Fog at the edge: experiences building an edge computing platform. In: 2018 IEEE International Conference on Edge Computing (EDGE) (2018)
6. Howard, A.G., Zhu, M., Chen, B., Kalenichenko, D., Wang, W., Weyand, T., Andreetto, M., Adam, H.: MobileNets: efficient convolutional neural networks for mobile vision applications (2017). Preprint arXiv:1704.04861
7. Huynh, L.N., Lee, Y., Balan, R.K.: DeepMon: mobile GPU-based deep learning framework for continuous vision applications. In: Proceedings of the 15th Annual International Conference on Mobile Systems, Applications, and Services, pp. 82–95. ACM, New York (2017)
8. Imamura, H., Sato, I., Sugiyama, M.: Analysis of minimax error rate for crowdsourcing and its application to worker clustering model (2018). Preprint arXiv:1802.04551
9. Kawano, M., Mikami, K., Yokoyama, S., Yonezawa, T., Nakazawa, J.: Road marking blur detection with drive recorder. In: 2017 IEEE International Conference on Big Data (Big Data), Dec 2017, pp. 4092–4097 (2017)
10. Lane, N.D., Bhattacharya, S., Georgiev, P., Forlivesi, C., Jiao, L., Qendro, L., Kawsar, F.: DeepX: a software accelerator for low-power deep learning inference on mobile devices. In: Proceedings of the 15th International Conference on Information Processing in Sensor Networks, p. 23. IEEE Press, Piscataway (2016)
11. Redmon, J., Divvala, S., Girshick, R., Farhadi, A.: You only look once: unified, real-time object detection. In: Proceedings of the IEEE Conference on Computer Vision and Pattern Recognition, pp. 779–788 (2016)
12. Shimodaira, H.: Improving predictive inference under covariate shift by weighting the log-likelihood function. J. Stat. Plan. Inference **90**(2), 227–244 (2000)
13. Sugiyama, M., Kawanabe, M.: Machine Learning in Non-stationary Environments: Introduction to Covariate Shift Adaptation. MIT Press, Cambridge (2012)
14. Toshniwal, A., Taneja, S., Shukla, A., Ramasamy, K., Patel, J.M., Kulkarni, S., Jackson, J., Gade, K., Fu, M., Donham, J., et al.: Storm@ twitter. In: Proceedings of the 2014 ACM SIGMOD International Conference on Management of Data, pp. 147–156. ACM, New York (2014)
15. Xin, D., Ma, L., Liu, J., Macke, S., Song, S., Parameswaran, A.: HELIX: accelerating human-in-the-loop machine learning. Proc. VLDB Endow. **11**(12), 1958–1961 (2018)
16. Yonezawa, T., Ito, T., Nakazawa, J., Tokuda, H.: SOXFire: a universal sensor network system for sharing social big sensor data in smart cities. In: Proceedings of the 2nd International Workshop on Smart, p. 2. ACM, New York (2016)
17. Zaharia, M., Das, T., Li, H., Shenker, S., Stoica, I.: Discretized streams: an efficient and fault-tolerant model for stream processing on large clusters. In: HotCloud '12 (2012)
18. Zheng, Y., Zhang, L., Xie, X., Ma, W.Y.: Mining interesting locations and travel sequences from gps trajectories. In: Proceedings of the 18th International Conference on World Wide Web, pp. 791–800. ACM, New York (2009)

Ethical and Social Aspects of Connected and Autonomous Vehicles: A Focus on Stakeholders' Responsibility and Customers' Willingness to Share Data

Steven Andorka and Kira Rambow-Hoeschele

1 Introduction

The Internet of Things (IoT) is becoming more and more present in today's world in many different parts of life. While a lot of ideas and technologies exist, many people do not seem to be ready or willing to adapt to technological changes and seem skeptical about them mainly due to privacy concerns. Furthermore, ethical concerns have been raised in the past, but have not been clearly answered. As many new connectivity features could and will potentially be integrated in electrified cars in the future, several ethical questions need to be answered, to be able to set the right focus on the most promising technologies.

When thinking of ethics, it is important to consider the functionality of certain connectivity features. Do they really improve the life of the passengers in terms of safety and convenience or do they rather distract them? Do they add value and how relevant are they at the cost of data? It has been found that both handheld as well as hands-free cell phone conversations are interlinked with decreasing driving performance [1].

However, with increasing levels of autonomous driving, this threat will decrease, and other applications will be in focus that improve productivity. In general, an increasing productivity will be of benefit for the whole society as in many Western countries, societies are aging. Making driving safer, more effective, and more efficient will bring benefits for all. For instance, IoT health applications might

S. Andorka
Glasgow Caledonian University, School of Engineering, Glasgow, UK
e-mail: steven@andorka.de

K. Rambow-Hoeschele (✉)
Robert Bosch GmbH, Headquarters, Automotive Strategy, Stuttgart, Germany
e-mail: kira.rambow-hoeschele@de.bosch.com

R. José et al. (eds.), *3rd EAI International Conference on IoT in Urban Space*, EAI/Springer Innovations in Communication and Computing,
https://doi.org/10.1007/978-3-030-28925-6_2

provide improvements for people that are unforeseen so far. Another example is a potentially increasing level of fairness due to higher transparency as drivers could profit from good driving behavior through benefits such as lower insurance premia. In addition, with a digital identity existent from car usage, public transport could be made fairer as fare evaders would be identified at any point in time. With cars recognizing their passengers, they could be disabled from the outside in case of theft or to catch a criminal [2–5].

One important connectivity feature in the future will be a digital assistant that is facilitating the life of passengers. This is opening up the question about the scope of virtual assistants and connectivity features in general with AI supporting them. Will vehicle intelligence outperform human skills and what will be the consequences? It is debated whether the intelligent car might become the first social robot that will interact with a large number of humans.

A very important issue is whether virtual assistants will only remain to assist people or whether they will take over decisions and tasks whenever possible. Literature defines two types of scenarios in which virtual assistants become "virtual butlers" only reacting to user requests or in which they will become proactive, self-aware, and autonomous as a "virtual companion." Recently a declaration has been published that recommends a limited or "constrained autonomy" due to the fear of intelligent robots replacing humans in a variety of fields. Following this argumentation, virtual assistants should only provide options to humans who will then have the power of decision-making [6–9].

Furthermore, with an increasing importance of IoT applications, it may become difficult to leave the network if wished as with time, dependability increases. Another important topic is IoT globalization, meaning that data will be exchanged across borders and potentially saved on a server in another country. With sensitive data crossing borders, potential problems could arise if, for instance, privacy and cyber security laws in that country change and, unexpectedly, third parties might receive access to private data.

This is also linked to the issue of identification. It has to be guaranteed that every device that is sending or receiving data has an individual identification that is unique. A data transfer to a wrong receiver could result in severe damage imagining, for example, that health data could be linked to a wrong personal profile. With gathering and digitizing data that has not been available before and is now exposed to potential threats at every time bears a high risk especially for highly sensitive data related to health or personal information. Whereas IoT solutions in the health sector could support decisions by doctors, wrong information included in a health profile, due to potential hacking attacks, creates new threats that have not existed before. Moreover, public and private life melt together and become accessible for a broader group of individuals than in the past leading to both positive and negative effects [10].

There is a broad consensus that car connectivity features have the capacity to provide major benefits for customers. Looking at social benefits, they create a time surplus when being in the car through providing driver freedom leading to increased productivity. Additionally, they can help reduce congestion and fatalities saving lives, time, and costs. With higher transparency and more data collected, it might be easier to measure the ethical footprint of every individual leading to an improvement of overall behavior of people. However, all of this comes at the price of data collection. According to studies, a large number of people are willing to provide some sort of data when receiving a fair value of benefits in return [11, 12]. However, one important question is what this fair value is and how it can be measured.

2 Thoughts on Ethics

2.1 Responsibility of Stakeholders

Although a large variety of connectivity features has already been existent in the past, many drivers resisted to buy or use them due to different reasons, such as not trusting or understanding the new technologies or by neglecting the value that they offer. However, with new technological developments coming up, it is projected that the value of connected car data and new business models emerging from the digitization of the car could be worth up to $1.5 trillion a year by 2030 [13]. With such an economic impact and many different parties involved and interested in connected car data, it is crucial to define certain ethical standards that need to be applied.

In literature, eight critical questions are mentioned that are necessary to be answered for an ethically sound connected car ecosystem. First, who owns the data? Second, who is able to control the data? Third, who can access the data? Fourth, is data integrity ensured, meaning that the data is accurate and consistent? Fifth, is data authenticity guaranteed, meaning that data is not manipulated? Sixth, is the data available at any moment when needed? Seventh, is the data processed in a confidential way? Eighth, are involved parties committed to keep ethical standards up? Especially the question of data access is of high importance as many parties such as governments, companies, employers, and insurances are interested in connected car data and personal data providing them power to make decisions upon people [14–19].

According to Arkin [20] autonomous technology is created faster than we are able to "(1) understand its implications, (2) interpret is with moral frameworks, and (3) create policy and legislation to govern its development and deployment." Often, technical progress evolves fast whereas legal and ethical aspects are not discussed in such a detailed way at an early stage when actually necessary. However, this is essential when it comes to sensitive data that is collected in the connected car as wrong usage could lead to harsh damage for individuals. Therefore, it is of high importance to establish ethical standards that find broad acceptance [21–24].

2.2 *Customers' Willingness to Share Data*

According to a study, personal data privacy does not seem to be of major concern for many customers. Many people share much data already today via their smartphones or social media. Only 25% of customers categorically refuse OEMs to use their driving and position data. Another study found that 82% of drivers surveyed worldwide are willing to share their vehicle data. Although different consent rates can be found depending on the study, a general tendency towards acceptance of connectivity features among society can be found. However, differences in the customer perception depending on the region exist. Chinese citizens, for example, tend to be more willing to share their data in comparison to Germans. In some regions, customers are even willing to share more data with their car's OEM than with their smartphone manufacturer. In general, driving-related connectivity features such as connected navigation or networked parking seem more important to customers than driving-unrelated ones such as e-mail or web browsing [25, 26].

On average, 71% of consumers consciously decide to grant certain applications access to data that they would not share with other applications. This number keeps increasing, as people tend to become more aware of potential benefits. Depending on the field of application, there are different levels of acceptance to share data. 37% would not like to use a car with many connectivity features due to privacy concerns. However, this differs on a regional level. Germans are above average with 51% being skeptical, whereas Chinese tend to be more willing to use them with only 21% being doubtful. On a global scale, 32% of car buyers would vote for a law that allows connected cars to enforce speed limits and would stop over speeding. However, large regional differences exist. In China 58% would support this initiative whereas in Germany only 15% would be willing to do so [11, 13, 26].

With new connectivity features coming up, new business models will evolve that will allow customers to use their data as a currency. One exemplary use case from the sharing segment would be a customer that receives free miles in the car in return for personal data that could be used for targeted advertising. In 2014, 35% of customers were willing to trade driving data in order to receive insurance benefits. Yet, when having the option to choose either chargeable services or free services in exchange for viewing advertisements or providing data, the customer view depends on the type of feature. Whereas people would rather pay for networked parking, they would tend to use free connected navigation. For connected navigation features, 70% of the customers in the USA are willing to share personal data, while the number drops to 58% for usage-based tolling and taxation features. One very extensive discrepancy could be identified for features using fitness and health data. In Germany, only 23% were willing to use those features in comparison to 43% in the USA, and 79% in China. Besides the regional differences, in general it could be observed that younger people and people who spend more time in their car tend to be more open towards car connectivity features and rather see them as exciting [11, 13, 26].

3 Conclusion and Future Work

With the technological developments in the field of connectivity features in the car many new opportunities are created. However, they do not come without any risks and threats. To ensure that future decisions will not only be reasonable from an economic point of view, but also from a moral one it is highly important to examine technological process in detail from an ethical perspective.

This thesis aimed to provide insights on the ethical perspective on connectivity features in the car. However, many questions have been left unanswered. As this is a constantly changing field of technology, new questions arise frequently. Interesting further research topics could be, for instance, the different customer perceptions on connectivity features depending on the region on a national or an international scale.

References

1. Rosenberger, R.: Driver distraction from dashboard and wearable interfaces: a case against connectivity. IEEE Technol. Soc. Mag. **34**(4), 88–99 (2015). https://doi.org/10.1109/MTS.2015.2494399
2. Buntz, B., Deloach, D.: AI and IoT could make a big splash in healthcare ... eventually. http://www.ioti.com/analytics/ai-and-iot-could-make-big-splash-healthcare-eventually (2017). Accessed 05 May 2018
3. Dimitrov, D.: Medical internet of things and big data in healthcare. Health Inf. Res. **22**(3), 156 (2016). https://doi.org/10.4258/hir.2016.22.3.156
4. Innovation Group.: Automotive future now report 2016 - preparing the automotive industry for the next decade. http://www.focusontransport.co.za/archive/images/Automotive_FNR_Layout_2016_digi.pdf (2016)
5. Mccluskey, B.: Connected cars – the security challenge. Eng. Technol. **12**(2), 54–57 (2017). https://doi.org/10.1049/et.2017.0205
6. Hassler, S.: Self-driving cars and trucks are on the move. IEEE Spectr. **54**(1), 6–6 (2017). https://doi.org/10.1109/MSPEC.2017.7802341
7. Holder, C., Khurana, V., Harrison, F., Jacobs, L.: Robotics and law: key legal and regulatory implications of the robotics age (part I of II). Comput. Law Secur. Rev. **32**(3), 383–402 (2016). https://doi.org/10.1016/j.clsr.2016.03.001
8. Knight, W.: New self-driving car tells pedestrians when it's safe to cross the street. https://www.technologyreview.com/s/602267/new-self-driving-car-tells-pedestrians-when-its-safe-to-cross-the-street/ (2016). Accessed 03 May 2018
9. Nass, C., Yen, C.: The Man Who Lied to His Laptop: What We Can Learn About Ourselves from Our Machines. Penguin Group, New York (2010)
10. AboBakr, A., Azer, M.: IoT ethics challenges and legal issues. In: 2017 12th International Conference on Computer Engineering and Systems (ICCES), pp. 233–237. IEEE (2017). https://doi.org/10.1109/ICCES.2017.8275309
11. McKinsey & Company: Will car users share their personal data? https://www.mckinsey.com/industries/automotive-and-assembly/our-insights/will-car-users-share-their-personal-data (2016). Accessed 02 May 2018
12. Strategy Analytics: Accelerating the future: the economic impact of the emerging passenger economy. https://newsroom.intel.com/newsroom/wp-content/uploads/sites/11/2017/05/passenger-economy.pdf (2017)

13. McKinsey & Company: Car data: paving the way to value-creating mobility - perspectives on a new automotive business model. https://www.mckinsey.de/files/mckinsey_car_data_march_2016.pdf (2017)
14. Boritz, J.: IS practitioners' views on core concepts of information integrity. Int. J. Account. Inf. Syst. **6**(4), 260–279 (2005). https://doi.org/10.1016/j.accinf.2005.07.001
15. Deloitte: Who owns the road? The IoT-connected car of today - and tomorrow. https://www2.deloitte.com/de/de/pages/manufacturing/articles/who-owns-the-road-iot.html (2015)
16. Jumelle, A., Ispas, I., Thuernmler, C., Mival, O., Kosta, E., Casla, P., et al.: Ethical assessment in e-Health. In: 2014 IEEE 16th International Conference on e-Health Networking, Applications and Services (Healthcom), pp. 262–268. IEEE (2014). https://doi.org/10.1109/HealthCom.2014.7001852
17. Poll, H.: Harris Poll AutoTECHCAST(SM) study: connected car high-tech features popular, but more marketing muscle needed to spur demand, adoption. https://www.prnewswire.com/news-releases/harris-poll-autotechcastsm-study-connected-car-high-tech-features-popular-but-more-marketing-muscle-needed-to-spur-demand-adoption-269031001.html (2014)
18. Sholla, S., Naaz, R., Chishti, M.: Incorporating ethics in Internet of Things (IoT) enabled connected smart healthcare. In: 2017 IEEE/ACM International Conference on Connected Health: Applications, Systems and Engineering Technologies (CHASE), pp. 262–263. IEEE (2017). https://doi.org/10.1109/CHASE.2017.93
19. The Economist: The world's most valuable resource is no longer oil, but data. The Economist. https://www.economist.com/news/leaders/21721656-data-economy-demands-new-approach-antitrust-rules-worlds-most-valuable-resource?fsrc=scn/fb/te/bl/ed/theworldsmostvaluableresourceisnolongeroilbutdataregulatingtheinternetgiants (2017)
20. Arkin, R.: Ethics and autonomous systems: perils and promises. Proc. IEEE. **104**(10), 1779–1781 (2016). https://doi.org/10.1109/JPROC.2016.2601162
21. Caron, X., Bosua, R., Maynard, S., Ahmad, A.: The Internet of Things (IoT) and its impact on individual privacy: an Australian perspective. Comput. Law Secur. Rev. **32**(1), 4–15 (2016). https://doi.org/10.1016/j.clsr.2015.12.001
22. Leggat, H.: Ethics and legal considerations in the Internet of Things (IoT). In: 2017 IEEE International Conference on Pervasive Computing and Communications Workshops (PerCom Workshops), pp. 611–611. IEEE (2017). https://doi.org/10.1109/PERCOMW.2017.7917632
23. Moor, J.: The nature, importance, and difficulty of machine ethics. IEEE Intell. Syst. **21**(4), 18–21 (2006, August). https://doi.org/10.1109/MIS.2006.80
24. O'Leary, D.: Ethics for big data and analytics. IEEE Intell. Syst. **31**(4), 1541–1672 (2016). https://doi.org/10.1109/MIS.2016.70
25. Accenture: AI is the new UI - experience above all. https://www.accenture.com/t20171005T065813Z__w__/us-en/_acnmedia/Accenture/next-gen-4/tech-vision-2017/pdf/Accenture-TV17-Trend-1.pdf#view=FitH (2017)
26. Habeck, A., et al.: Connected Car, Automotive Value Chain Unbound, 1–50. McKinsey & Company (2014)

Evaluating Street Networks for Predictive Policing

Julio Cezar De Melo Borges, Nelio Cacho, Long Wang, and Michael Beigl

1 Introduction

Much of past research in environmental criminology explores how the architecture of street network has an impact on crime activity, as for example by suggesting that there are fewer crimes on cul-de-sacs and dead-end streets than on through streets [8]. On the other hand, it has also been explored how street networks can influence the movement patterns of criminals and how these patterns can affect the decision process in target selection [1]. These studies have influenced urban street design and planning for years.

Urban designers modify traffic flows, based on the hypothesis that traffic flow can have substantial impacts on a city's crime patterns [4]. They are often in favor of lower densities and less connected places, since reduced accessibility decreases the availability of escape routes to criminals while increasing the level of street control that residents can exert as there are less passersby. However, extensive studies considering many several aspects of the street network, scalable and applicable to several crime historical records of different regions are rather rare [12].

In this work, we tackle the problem of automatically extracting several properties (features) of street networks, correlating this information with crime patterns. We leverage foundations from graph theory in order to model street networks as graphs with edges representing street segments and nodes as intersections between them. We them extract from this representation several features from the modeled street

J. C. De Melo Borges (✉) · L. Wang · M. Beigl
TECO, Institute of Telematics, Karlsruhe Institute of Technology (KIT), Karlsruhe, Germany
e-mail: borges@teco.edu; wanglong@teco.edu; beigl@teco.edu

N. Cacho
Federal University of Rio Grande do Norte (UFRN), Natal, Brazil
e-mail: neliocacho@dimap.ufrn.br

R. José et al. (eds.), *3rd EAI International Conference on IoT in Urban Space*,
EAI/Springer Innovations in Communication and Computing,
https://doi.org/10.1007/978-3-030-28925-6_3

network. We propose using these street network features for building analytics models in different urban scenarios. In this paper specifically, we correlate this information with crime patterns in the city of San Francisco, delivering and discussing an exemplary use-case in a predictive model for Crime Hotspot Detection based on the feature of the street network.

This paper is organized as follows. Section 2 presents the related work in extracting features from the urban space and applying these features in urban computing scenarios. Section 3 describes the extraction of street network features for the city of San Francisco. We give a detailed explanation of these features, researching their spatial autocorrelation in Sect. 4. Section 4 also delivers an exemplary use-case for using such features in the area of Crime Hotspot Detection. Section 5 concludes this paper and describes on-going and future work.

2 Related Work

Extracting features from the urban space has been gaining increasingly importance for building pattern recognition models in a variety or urban scenarios, specially in predictive policing [13]. Predictive Policing is an area of Urban Computing which leverages analytical techniques to identify potential criminal activity. Specifically, Predictive Policing focuses on identifying locations at increased risk of crime. It is a tool to provide insights to police strategists, e.g., for police patrol planning patrol in order to make the best use of resources or to have the greatest chance of deterring or preventing future crimes [6].

Venerandi et al. proposed extracting a set of descriptors of the urban environment from a mix of open datasets, proposing a set of significant features which were interpreted to provide a plausible description of criminal activity in London area [12]. Venerandi et al. showed that a street network feature, namely the density of dead-ends (streets which end with no intersections) showed the highest importance value for his task. In this work, we confirm this correlation but show how this feature is spatially correlated.

Borges et al. evaluated a series or spatio-temporal and urban features for the task of Crime Hotspot Detection for the cities of Natal, Brazil and San Francisco, USA, identifying the locality importance of such features (i.e., features which have importance in one area but not in another) [3]. Borges et al. showed that specially street network features such as the ratio from Major to Minor Roads or the Length of Pedestrian Streets in an area highly (and universally) correlate to high criminal activities. However, they just considered a few street network features directly extractable from the urban space and did not consider modeling the street network as a graph for extracting further graph-theoretical features from the street net as this work proposes.

Both the works of Venerandi et al. and Borges et al. highlighted the importance of Street Network features for such crime/urban analytics tasks. This work extends previous results in this area.

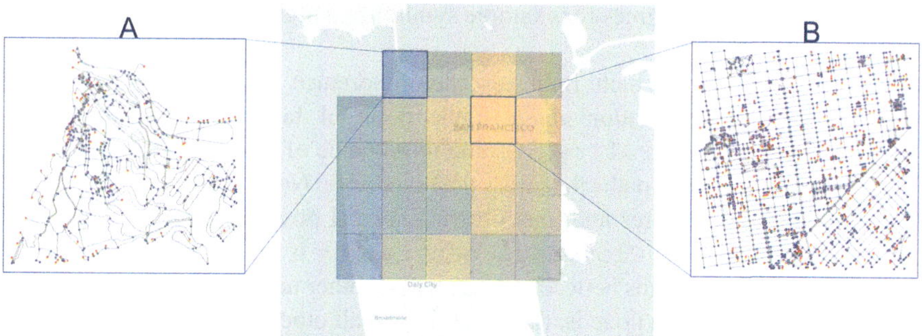

Fig. 1 Spatial division of San Francisco by grids of 3 km size highlighting spots of theft criminal activities (low to high = blue to orange). The street network in cold and hot spots differ in their structure and topology. Cul-de-sacs depicted red

3 Street Network Features

Core contribution of this paper lies on modeling the street net into an undirected graph for street network feature extraction, calculating various metrics on top of graph topological measures. In our representation, the street net is modeled into a graph where nodes represent street segments and edges represent intersections as proposed by Porta et al. [10]. Figure 1 shows this graph representation for regions of the city of San Francisco, cul-de-sacs being highlighted red.

We then extracted several metrics from topology and structure of the modeled street network. `Intersection count` is the count of nodes with crossing streets—i.e., with a degree higher than 1—in an area. Streets with a high intersection density offer more possibilities for escape routes [5].

`Dead-ends` or cul-de-sacs are then nodes with degree 1, i.e., not having any intersection with another street. It is given as absolute and relative (proportion per total nodes) number per area. The proportion of dead-ends in an area limits possible escape routes [5].

`Node count` is the total number of nodes per Area, disregarding if they are dead-ends or intersections. `Edge count` is the total number of edges, differing from `Street Segments Count`, the number of edges—counted in all directions—providing information on how fine-grained the street network is.

The `Mean edge length` depicts the average street length (in meters) of the modeled street network. `Street Segments length` depicts street length in undirected representation of network.

`Average Circuity` is the ratio of the street length by the linear distance between two linked nodes in the graph. It represents how "directed" connected (or curvy) street segments are.

Several topological measures of the graph model of the street network can reveal some information about it. `Average node degree` depicts how well the nodes are connected: lower degrees for example symbolize a high proportion of dead-ends and less connectivity.

Centrality metrics which indicates the importance of nodes in a network reveal important information about single point of failure in possible routes. `Betweenness centrality` gives the importance of a node in the network by the number of (shortest) paths that cross the node/edge. Nodes with high importance are single point of failures and may obstruct several possible routes in an area as a measure of resilience. `Closeness centrality` is the reciprocal sum of the length of the shortest paths between a node and all other nodes in the street network. Thus, the more central a node is, the closer it is to all other nodes.

Several other graph measures can also be extracted from the modeled street network for serving as street network feature such as the `Clustering Coefficient` and `PageRanks`. We refer to [2, 11] for a detailed explanation of these and further measures.

4 Feature Analysis

To carry out this study, we leverage a historical database of crime incidents for the city of San Francisco between 2003 and 2015. We focused only on robbery-related crime categories in this study, which we assume to be more dependent on escape routes and therefore on the street network.

For carrying out feature correlation and importance analysis, we divided the city of San Francisco into equidistant grids of 3 km in size, resulting into a partitioning of the space in 24 subdivisions (see Fig. 1).

For several use-cases in Urban Computing and Predictive Policing, data-based models require the absolute independence of observations [12]. However, the street network features proposed in this work might show spatial autocorrelation, i.e., the tendency of nearby observations to be correlated to one another. We perform the well-known geostatistical Moran's test to rule such features out [9]. The Moran's Index delivers the strength of the spatial dependency of the tested variable and is shown in Table 1.

As a measure of feature importance we leverage the Mutual Information (MI) metric [7], which measures the dependency between the crime incidents and the proposed features. We calculate this measure for the regions of San Francisco illustrated in Fig. 1. It is equal to zero if and only if two random variables are independent, and higher values mean higher dependency.

Table 1 summarizes the analysis for spatial autocorrelation and feature importance. Dead-ends are by far the most important street network feature related to robbery crimes in San Francisco. But this feature is also spatially correlated in San Francisco, which has regions with similar dead-end amounts in nearby regions. The count of street segments on the other site is also highly important and not spatially correlated, making it a suitable feature for pattern recognitions models based on this dataset.

Table 1 Results of Moran's Test and values for Mutual Information (MI) for the proposed Street Network Features and crime incidents for the city of San Francisco

Feature	Moran's Index	p-value < 0.05	MI
deadends	0.291851	True	0.466
street_segments_count	0.020423	False	0.368
nodes_count	0.065096	False	0.332
closeness_centrality_avg	0.052619	False	0.328
intersection_count	0.013863	False	0.306
betweenness_centrality_avg	0.029080	False	0.254
clustering_coefficient_avg	−0.063050	False	0.241
street_length_total	0.047213	False	0.230
degree_centrality_avg	0.035916	False	0.224
edges_count	0.004315	False	0.223
pageranks	0.008897	False	0.201
edge_length_avg	0.313867	True	0.114
edge_length_total	0.090583	False	0.104
circuity_avg	−0.115039	False	0.073
avg_neighbor_degree_avg	0.209038	True	0.063
deadend_proportion	0.287994	True	0.010

Use-Case: Crime Hot-Spot Detection In order test the spatially independent features in a concrete use-case scenario, we train a machine learning model (Random Forest) for the task of Crime Hotspot Detection as described in [3] and evaluate its performance based on a Leave-One-Out validation. Regions with high crime incidents (>80% quantile) are deemed a positive class, i.e., a Hot Spot. This model achieves a F1-Score of 0.60 for the positive class and a F1-Score of 0.89 for the negative class, thus achieving an average of **0.83** for the F1-Score metric normalized by class occurrence. These results deliver indication that these features can be suitable for Urban Computing use-cases in general and Predictive Policing in particular.

5 Conclusion

In this work, we proposed to make use of features extracted from street networks for pattern recognition in Urban Computing and specially in Predictive Policing scenarios. We showed the relationship between the street network and robbery incidents for the city of San Francisco, evaluating how street network features correlate with these crime incidents. Additionally, we also showed that some of these features are spatially correlated, being thus spatially dependent and therefore not suitable for generalizable machine learning models. We further exemplarily showed the application of street network features in a Crime Hotspot Detection model, achieving promising results. In the future, we plan to evaluate such features in an even larger dataset corpus and in other urban computing scenarios.

References

1. Beavon, D.J., Brantingham, P.L., Brantingham, P.J.: The influence of street networks on the patterning of property offenses. Crime Prev. Stud. **2**, 115–148 (1994)
2. Boeing, G.: OSMnx: New methods for acquiring, constructing, analyzing, and visualizing complex street networks. Comput. Environ. Urban Syst. **65**, 126–139 (2017)
3. Borges, J., Ziehr, D., Beigl, M., Cacho, N., Martins, A., Sudrich, S., Abt, S., Frey, P., Knapp, T., Etter, M., et al.: Feature engineering for crime hotspot detection. In: 2017 IEEE SmartWorld, Ubiquitous Intelligence & Computing, Advanced & Trusted Computed, Scalable Computing & Communications, Cloud & Big Data Computing, Internet of People and Smart City Innovation (SmartWorld/SCALCOM/UIC/ATC/CBDCom/IOP/SCI). IEEE, Piscataway (2017)
4. Clarke, R.V.: Situational crime prevention. Crime Just. **19**, 91–150 (1995)
5. Hillier, B., Sahbaz, O.: High resolution analysis of crime patterns in urban street networks. Techne Press, Amsterdam (2005)
6. Junior, A.A., Cacho, N., Thome, A.C., Medeiros, A., Borges, J.: A predictive policing application to support patrol planning in smart cities. In: 2017 International Smart Cities Conference (ISC2) (Sept 2017). https://doi.org/10.1109/ISC2.2017.8090817
7. Kraskov, A., Stögbauer, H., Grassberger, P.: Estimating mutual information. Phys. Rev. E **69**(6), 066138 (2004)
8. Nutter, J., Bevis, C., Minnesota Crime Prevention Ctr, United States of America: Changing street layouts to reduce residential burglary. Minnesota Crime Prevention Center, Minneapolis (1977)
9. Ord, J.K.: Spatial processes. Wiley StatsRef: Statistics Reference Online (2014)
10. Porta, S., Crucitti, P., Latora, V.: The network analysis of urban streets: a primal approach. Environ. Plan. B Plan. Des. **33**(5), 705–725 (2006)
11. Trudeau, R.J.: Introduction to Graph Theory. Courier Corporation, North Chelmsford (2013)
12. Venerandi, A., Quattrone, G., Capra, L.: Guns of brixton: which london neighborhoods host gang activity? In: Proceedings of the Second International Conference on IoT in Urban Space. pp. 22–28. ACM, New York (2016)
13. Zheng, Y., Capra, L., Wolfson, O., Yang, H.: Urban computing: concepts, methodologies, and applications. ACM Trans. Intell. Syst. Technol. **5**(3), 38 (2014)

Part II
Design Challenges in Urban IoT

Challenges in Using IoT in Public Spaces

Ulf Hedestig, Daniel Skog, and Mikael Söderström

1 Introduction: Citizens as Co-Constructors

With the arise of globalization and digitalization new perspectives on public administration have emerged such as New Public Service [1, 2], Public Value Management [3], New Public Governance [4], and Digital Era Governance [5]. These perspectives have a mutual starting-point, the "reasonable man," which is based on dialogue and deliberation. In this view, the citizen often is regarded as a problem solver and co-creator engaged in creating what is valuable and good for the public together with the public sector. Other consequences of the new perspectives concern the shift from an inside-out approach of offering service to focusing on service or citizen journeys, from plan-driven projects towards more agile ways of working, and realization of a do-it-yourself strategy with self-service opportunities [6].

Several of these new perspectives seem to have the concept of co-production in common [7], or co-creation as it sometimes is called, meaning that the public sector engages citizens not only to provide opinions on the services it delivers, but also to take part in the design of new services, the redesign of existing services and the actual execution of services, partly or fully. A quite recent way of establishing co-production in the public sector has been to apply information technologies into active or passive settings. One example of an active setting is crowdsourcing [8], and one example of a passive setting is social media monitoring [9].

However, the way crowdsourcing has been implemented in public sector is often associated with difficulties, and some of them are also related to social media monitoring. Perhaps the most important issue is related to the nature of the crowd.

U. Hedestig · D. Skog (✉) · M. Söderström
Umeå University, Umeå, Sweden
e-mail: ulf.hedestig@umu.se; daniel.skog@umu.se; mikael.soderstrom@umu.se

R. José et al. (eds.), *3rd EAI International Conference on IoT in Urban Space*, EAI/Springer Innovations in Communication and Computing,
https://doi.org/10.1007/978-3-030-28925-6_4

The question of who forms the group, how it is organized, and what task the crowd is involved in effects the outcome of the crowd's performance. Challenges that need to be managed include, for instance, the existence of a sufficient diverse and knowledgeable active crowd, the task and the crowdsourcing process that the crowd is supposed to participate in, and the channel, or the media, in which the crowdsourcing occurs.

To overcome some of the difficulties associated with crowdsourcing and social media monitoring we propose that an alternative or complement may be to use Internet of Things (IoT) technology to gather data regarding citizens' everyday behavior and the actions they perform in the real world. By doing this we are trying to investigate the research question whether or not sensor data can be used as an input to data-driven business development. You might say that this is an IoT approach to citizen sourcing that hopefully will create improved conditions for business development, and innovative citizen sourcing in public sector organizations. Janssen et al. [10] present similar ideas of data-driven innovation in the public sector but in a big and open linked data view that is more comprehensive than the IoT view of this article. The IoT approach is at the moment tested in a local government in northern Sweden, where we have set up a Low Power Wide Area (LoRa) Network specifically designed for wireless battery-operated sensors.

2 The Context: A LoRa Network in a Scarcely Populated Area

2.1 Research Site

The local government in which we test the IoT approach has a population less than 5000 inhabitants. The LoRa infrastructure that we have implemented has a star topology with gateways as bridges between sensors and a central server [11], LoRa was chosen since it is an open and widely spread platform, including an open, non-profit association sharing their experiences.

By placing two gateways in central locations, the network covers a large part of the central city area. Each gateway is expected to cover a radius of three kilometers (Fig. 1). The reason for also including the outskirts of the town is to cover leisure areas such as cross-country skiing trails, cross-training tracks, barbecue areas, and areas for swimming and hiking.

In the initial phase, the placing of sensors was based on some of the current challenges the business development team at the local government has on their desk. Examples of issues are: (a) the city has too many pedestrian crossings, and other attempts on deciding which crossings that should be eliminated have failed. Questionnaires and focus group interviews have not given any clear answers regarding the citizens' use of these crossings, (b) the city would like to know more

Fig. 1 Two LoRa gateways covering the central city

about when and how much their outdoor multi-sport arena is used, and (c) how often cleaners should clean public toilets and empty trash cans/baskets.

To further explore what kind of sensors that could be implemented, the business development team performed focus group interviews with different target groups during fall 2017. Target groups can either be by a specific category, e.g., non-profit organizations or property owners, or by age. The aim of these focus groups has been to investigate what information/service the citizens would like to have from the local government and discuss which channels they prefer for interaction. These meetings have uncovered requests as for instance, "knowing how many people that have joined different public areas such as the gym or the public bath," or "knowing snow and air temperature for preparing cross-country skis in an appropriate way."

The initial investigation shows that citizens' need of information is much bigger than what the local government is currently providing. Several of the suggestions are opposite of what we can find in large cities that have adopted the smart city concept, where much of the information is about avoiding queuing, traffic jams, and crowds. In our case with a sparsely populated setting, the citizens would like to know where other people are at the moment, especially the young population. Even if they can use social media to interact, some of the population such as young immigrants and asylum seekers has not yet been incorporated and obtained access to these local digital communities.

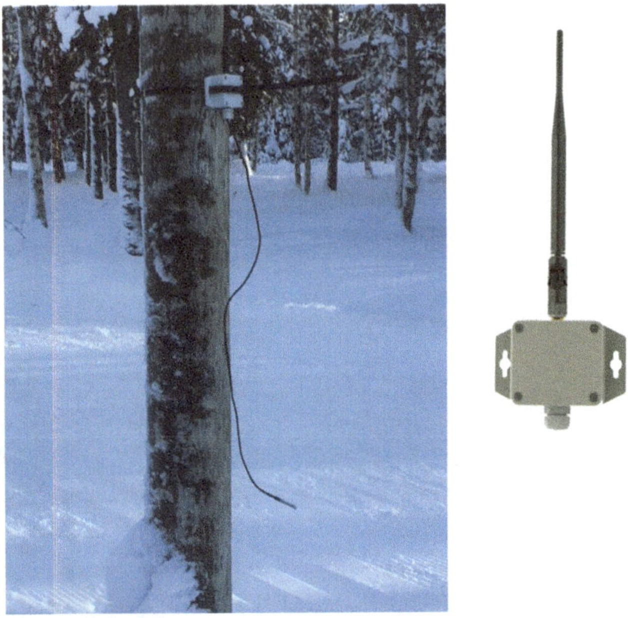

Fig. 2 Sensors measuring temperature and humidity was installed at the cross-country skiing track

Further, the focus group interviews also show that different target groups are familiar with different social media platforms. For the young population the majority use Snapchat in their communication. On the other hand, middle aged and older people prefer Facebook or Instagram. This can influence the choice of platform for visualization of sensor data.

2.2 *Initial Responses on Sensor Data*

The first sensors were deployed during the first quarter of 2018, based on current challenges of the business development team and input from focus group interviews. The placing emanates mainly from citizens' interest of sport and health, with the purpose of delivering data to enhance services. Two sensors were therefore placed at a cross-country skiing trail, one at the start/finish and the second one on the opposite side of the track, a couple of kilometers from the start (Fig. 2). Both sensors are measuring the snow temperature and humidity, which are of importance when the skiers select appropriate ski wax based on the current weather conditions.

The local ski club arranged a cross-country ski competition during the same time period and wanted to use the sensor data at once, so we provided them with a link to the server where sensor data were stored. Then the ski club put the link to the raw

#	Time	Temp	Humidity	Vdd	Exttemp	RSSI	SNR	SF	GW	NO
1140	2018-03-18 10:17:26	3.1	13	3.408	3.4	-108	7	7	C49300FFFF0EDD88	1290
1141	2018-03-18 09:57:26	2.7	13	3.403	3.5	-112	-2.5	7	C49300FFFF0EDD88	1289
1142	2018-03-18 09:37:26	1.7	13	3.396	2.8	-114	1.5	7	C49300FFFF0EDD88	1288
1143	2018-03-18 09:17:27	1.3	13	3.392	1.6	-105	7.3	7	C49300FFFF0EDD88	1287
1144	2018-03-18 08:57:27	1	13	3.387	1.5	-107	7.3	7	C49300FFFF0EDD88	1286
#	Time	Temp	Humidity	Vdd	Exttemp	RSSI	SNR	SF	GW	NO
1143	2018-03-18 11:08:36	2.6	14	3.4	3	-77	8.8	7	C49300FFFF0EDD88	1289
1144	2018-03-18 10:28:36	2.1	14	3.395	2.8	-81	10.8	7	C49300FFFF0EDD88	1287
1145	2018-03-18 10:08:36	1.8	14	3.391	2.6	-76	8.5	7	C49300FFFF0EDD88	1286
1146	2018-03-18 09:48:36	1.3	14	3.386	2.2	-78	9.3	7	C49300FFFF0EDD88	1285
1147	2018-03-18 09:28:36	1	14	3.384	1.6	-78	8	7	C49300FFFF0EDD88	1284

Fig. 3 Sensor data from two positions at the cross-country skiing trail

data from the sensors on their website, without any re-configuration or reflections on interface design (Fig. 3).

As shown in Fig. 3, data was received every 20 minutes, and beside temperature and humidity it also included lots of irrelevant and internal information, such as the identifier number of the sensor, the temperature inside the sensor, signal strength, etc.

After the ski competition a few responses from the citizens were sent to the local government by SMS or through social media posts. Two examples of these are:

> Great! Thanks for thinking about us. (Informant 1, March 19, 2018)

> To me it was a big difference; I usually get up at 5.30 AM to read the temperature, and then update about every half hour. I didn't need to do that. Also, people usually call me and ask for the temperature, but during this weekend no one called me. (Informant 2, March 19, 2018)

Despite the lack of an appropriate interface, people interested in the snow temperature apparently did not have any problems to interpret the data tables offered by the ski club, and also found the available data to be very useful.

A second case showing the sensors ability to be a key driver for business development, and also their ability to generate value for different stakeholder groups, was when the business development team placed sensors at the local public bath. Sensors were placed both in the swimming pool measuring the water temperature and on the doors to the dressing rooms (Fig. 4). The purpose was to give the citizens information about the current temperature in the water and the quantity of visitors at the bath.

Citizens interested in swimming could recognize differences in temperature during the week, where Wednesdays was the day when the swimming pool was warmest. The reason for this is that it is physiotherapy and baby swim on Wednesdays and it requires a temperature of 32 °C. The rest of the week the

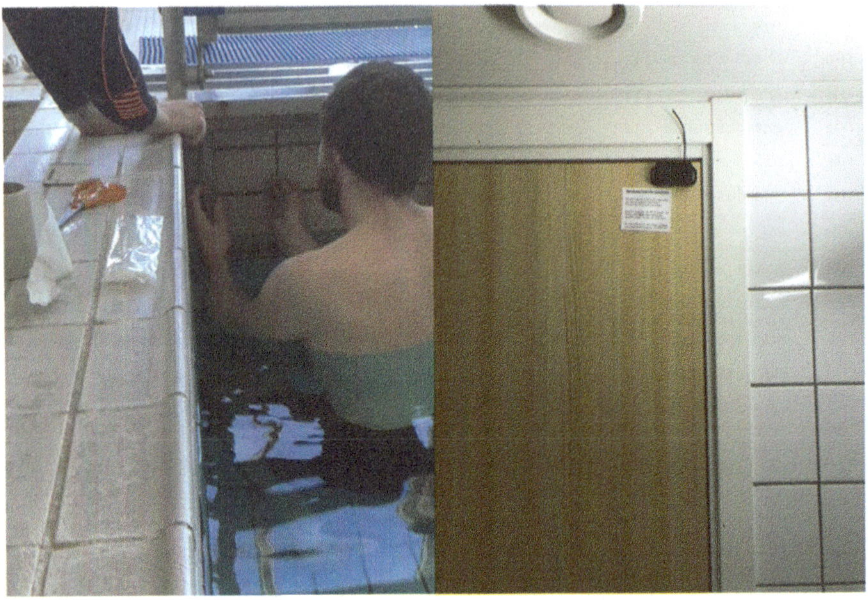

Fig. 4 Installing sensors at the public bath

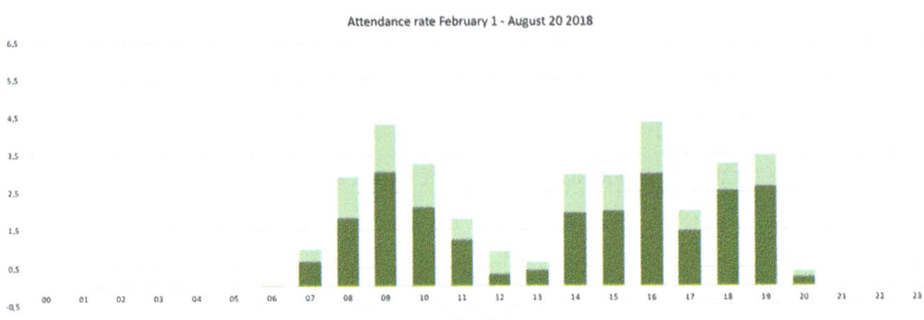

Fig. 5 Attendance rate at the public bath February to August 2018

temperature slowly dropped to become below 27° on Mondays. An unexpected advantage with the sensors was discovered when the local swimming club applied to the local government for extended training times at the public bath.

They presented a number of suggestions of times that was appropriate for themselves, but the head of the local government realized that the local government had sensor data on visitors and decided to use them as a decision basis. She made an analysis based on 6 months of sensor data (Fig. 5), which later on was used by the local politicians when deciding on the swimming club's training times.

A third case, involving several stakeholder groups, concerns the school environment. The first set of sensors was placed on the restroom doors at the school in the local government. Our purpose was to measure how frequently the restrooms were

Fig. 6 Example of destroyed sensors

used, and based on these data estimate how often each of them had to be cleaned. However, the business development team discovered that the restrooms in the school main building were almost not used at all. Instead, the children used the rooms in the next building. In a follow-up interview with the children, they explained that outside the restrooms in the main building, there was an open space where children spent their time during breaks, chatting, playing table tennis, etc. This space was so disturbing, with for instance people kicking on the doors, that it made them to avoid the restrooms. With these data at hand, the school management started to discuss changes in the school environment to improve the current situation and make the facilities more available.

A second set of sensors is currently being placed in classrooms, measuring the noise level. The aim is to try to accomplish a low noise level by informing different stakeholder groups (parents, teachers, pupils, principals, etc.) about the sound conditions in the classrooms. A meeting with all teachers and the school management showed that all of them are highly motivated to use sound sensors, but they decided to avoid to visualize online data from the sensors in the classrooms. The reason is that earlier attempts with other types of artifacts, visualizing noise level by a color, or a smiley, have resulted in pupils wanting to test the equipment by shouting loudly. Instead, the first test will include visualizations in form of a weekly report showing historical sound data as a basis for discussion among different stakeholder groups.

A concern with using sensors in public spaces was discovered when entered our fourth case in which we installed sensors measuring movements at a newly built outdoor multi-sport arena. The arena is roughly 120 × 75 m and contains a soccer/basket field, a playground, running tracks, a skateboard area, an outdoor gym, etc. The aim was to create a heat map of the area to analyze when the arena was used and which facilities that were most used. To our surprise, most of our sensors were destroyed in a few weeks, and parts of them lied scattered around the arena (Fig. 6).

The reason for this is unclear as we have not yet done any field study at the arena, but one possible explanation may be the design of the sensors. The sensors we use are encapsulated in black and white boxes and might look suspicious to people, in this case, probably quite young people. This problem has been addressed in earlier research, see for instance Jetter et al. [12], discussion of how people wary of being observed in communal and urban spaces. Therefore, we believe that we need to think quite a lot more on how to design sensors that either inform the citizen of the purpose of the measurement in a better way or are so discreet that they are not noticed.

3 Some Early Findings Regarding Use of IoT in Business Development

The four cases discussed above allow us to explore how IoT technologies can be used to support business development that hopefully also creates public value. Working together with the local government, we have been able to implement an IoT-network and test different types of sensors that are generating data that should be of interest for different stakeholder groups. Our purpose has been to investigate the potential of using IoT as a driver for business development. This is on-going research, but even at this early stage we have noticed some issues that are of importance for design and use of IoT in public spaces that are scarcely populated.

3.1 The Importance of Context

As can be seen in chapter two we have had different outcomes from our four cases. Perhaps the most surprising outcome is that the sensors were destroyed in one case but not in the other three. This may partly be due to privacy issues and the relation between privacy and context. The last years the privacy issue has become much more complicated in relation to the increasing amount of data being recorded by sensors in different contexts. Nissenbaum [13], and Baldini et al. [14], emphasize that privacy protection also should be related to the context in which the user is situated. The concept of context is here understood as consisting of an intertwined relationship between technology, practices, industry, and socio-cultural features, and since IoT can be used everywhere (at office, in public spaces or at home), context become a crucial aspect that has to be considered. A public space outside a restroom in a school is a different context than the multi-sport arena. The school case shows that pupils seem to regard the sensors as something that is in accordance with the school environment, in the same sense as equipment that controls ventilation and lightning. The multi-sport arena, on the other hand, is an outdoor open space available 24/7, and also a space where no authorities are visible for most of the time. The visitors at the arena might regard the space as something that they own and fully control. Black and white boxes without any information about what they are intended for may look suspicious in this context. The implementation of IoT sensors in public spaces with the aim of creating smart cities or supporting business development need to acknowledge that such contexts are complex, dynamic, and messy and cannot be designed with a one-size-fits all policy [15]. Hence, context has to be considered when designing IoT systems for public places. One way to do this may be to emanate from a systems approach and use rich pictures showing different stakeholder group's perspectives on a specific context [16].

A lack of citizens' awareness of IoT systems in public spaces can create problems in getting feedback from citizens [17]. If citizens are not informed of the purpose of sensor use they can become suspicious and act in different ways than their usual behavior, or even refuse to enter a specific space. In our cases one reason to why some of the sensors where destroyed may be that we put a note on every sensor in the school informing of the purpose and what the sensor was measuring, but not on the outdoor sensors at the multi-sport arena.

Awareness of sensors in different contexts can also result in diverse interpretations [17]. Rather neutral data such as humidity, temperature, dust, and energy can be unproblematic in public spaces, but could be regarded sensitive in private spaces such as your home where you probably do not want to measure the amount of dust.

3.2 Data-Driven Analytics from IoT Require Qualitative Analyses as Well

After having spent some time analyzing sensor data we realized that analyses of these kind of data in some cases also need to be complemented with qualitative analyses, since changes in people's movements and behaviors are difficult to interpret. In three of our cases we could see that sensor data was not enough to understand some situations or some behaviors. A sensor placed on a door measuring movement could become extremely complex to analyze. In such cases, did a registration of a movement mean that a person entered the next room through the door, or was the person just opening the door and did not enter the room? Also, the data registered did not measure the number of individuals possibly passing the door.

Given our purpose of creating data-driven business development which in turn leads to the development of improved and new public services, it is important to include ethnographical inspired methods such as observations or focus group interviews to get a deeper understanding of citizens behavior and movements in a public space equipped with sensors.

3.3 Suspicious Black Boxes in Some Cases Need to be Re-Designed

One reason for the different outcome of our four cases may be the physical design of the sensors. Most of the sensors used in public spaces are designed for other types of purposes, such as monitoring production processes and traffic flows, and controlling signal systems, ventilation, drain, energy consumption, and water consumption. In most cases those sensors have been developed with functionality in foreground and not with design in mind. They are black and white boxes that do not reveal any information what they do and which purpose they have. This circumstance can be of importance regarding citizens' perceptions and attitudes towards being sensed by IoT.

Currently there exist attempts to create other types of sensors with a more esthetic design, e.g., the Fukushima wheel, the Air Quality egg, and the IoT Egg. However, at the moment these types of sensors are very few in relation to the number of available sensors on the market.

3.4 Visualizations of Sensor Data Should be Related Towards Different Stakeholder Groups

We believe that an important factor for successful use of IoT in business development is to have a good strategy for visualization of sensor data. One of our coming key issues is to experiment with visualizations of the data received from the sensors. The aim is to find ways of creating value-added information and services to the citizens, for instance by publishing visualizations in different social media, which at the same time create a platform for discussing business development and improvements of public services. Hence, a future challenge is how to integrate different communication channels, such as social media, mobile apps, and web sites, in the dialogue and also how these can be integrated in the methods and tools used for innovation and business development in the public sector.

Our cases show that different stakeholder groups have different motives and requirements, which affect the choice of visualization method. One example of a method is shared public displays showing sensor data when individuals enter a particular place. Another option is to let individuals receive sensor data in a mobile app. Further, the locus of received data can also be of interest, where individual seems to prefer remote data access before local access [17]. To receive data about certain conditions at a remote place is similar to getting data from sensors at home when working. Local access means that when you enter a place you get actual data on a display from that place.

Other types of visualizations of sensor data can be important for certain stakeholder groups, for instance, governmental officials or politicians. Historical data over a longer period of time can provide support for business development of governmental services and help deliver better services to the citizens.

A strategy for visualization of sensor data has to consider that the solution would be designed for non-expert users, that is, different stakeholder groups among the citizens. A key issue here is if the sensor data only should provide actual plain data from sensors or provide data as a trigger for engagement and of certain behavior of citizens [15].

4 Conclusions

The idea of calm technology [18], which means that technology fades into the background or periphery, or Laurel's [19], concept of possibility, probability, and necessity, becomes a challenge when we talk about sensors. During the last decades a lot of focus has been on making technology more adapted to individual's preferences and invisible to the environment. To apply the same approach towards the emergence of IoT might not to be straightforward. The discussions of hiding sensors or to put them in foreground and clearly show them have as we can see in the above cases impact on people's actions and behavior. The physical design of sensors is still in its infancy, which may lead to unclear perceptions and understandings of what they do and for what purpose they do what they do.

One way of understanding sensors in different contexts and how they can be designed is to utilize the concept of affordances. Affordances originate from Gibson's ecological psychology [20], and were introduced into human–computer interaction by Norman [21]. Gibson's approach claims that affordances towards an artifact is dependent on experiences and culture in a specific context. Affordance for an individual in a specific environment depends therefore on the perceptual abilities of the individual [22]. An experienced HVAC technician might perceive an air sensor in an office building as affording a system for efficient air conditioning. For an office worker in the same building the same sensor can be perceived as something that is measuring temperature or it may even become "invisible" since you have no experience of sensors and what roles they play in that particular context. These different perceptions will determine both possible actions and which action capabilities that exist in a specific context.

Breakdowns can arise when there is a mismatch between perceived and actual affordances in a context. They can be seen as hidden affordances, that is, there exist possibilities for action but they are not perceived by the persons in the context. In our case it can be explained as citizens in a certain context that do not understand what the black box (the sensor) is and what purpose it has, and can therefore not perceive any possible actions. Another example is false affordance, which is a situation when persons in a certain context perceive nonexistent possibilities for action. This can perhaps explain why the sensors at the multi-sport arena were destroyed. It may be that the persons at the multi-sport arena interpreted the black boxes as something connected to surveillance an acted according to that, although the sensors do not possess that kind of functionality at all [23].

The concept of affordance can be useful to understand how we design and implement sensors in different contexts, that is, at home, at workplace, and in public spaces. The same sensors will be perceived and interpreted differently in different contexts, which will result in different actions and action capabilities. To fully take advantage of such an approach, we need further research based on ethnographical field observations focusing on people's behavior and actions in different public spaces equipped with sensors.

Acknowledgements We would like to thank Interreg Botnia-Atlantica 2014-2020 (grant no. 20201055) and Sweden's Innovation Agency (grant no. 2017-02133) for financially supporting the research project discussed in this paper. We would also like to thank Katarzyna Wikström for providing us with the statistics in Fig. 5.

References

1. Denhardt, R.B., Denhardt, J.V.: The new public service: serving rather than steering. Public Adm. Rev. **60**(6), 549–559 (2000)
2. Denhardt, J.V., Denhardt, R.B.: The new public service revisited. Public Adm. Rev. **75**(5), 664–672 (2015)
3. Stoker, G.: Public value management: a new narrative for networked governance? Am. Rev. Public Adm. **36**(1), 41–57 (2006)
4. Osborne, S.P.: The New Public Governance? Emerging Perspectives on the Theory and Practice of Public Governance. Routledge, London (2010)
5. Margetts, H., Dunleavy, P.: The second wave of digital-era governance: a quasi-paradigm for government on the Web. Philos. Trans. R. Soc. A Math. Phys. Eng. Sci. **371**(1987), 1–17 (2013)
6. Danneels, L., Viaene, S.: How to move towards digital era governance: the case of VDAB. Paper presented at the Proceedings of the 16th Annual International Conference on Digital Government Research, New York, NY (2015)
7. Linders, D.: From e-government to we-government: defining a typology for citizen coproduction in the age of social media. Gov. Inf. Q. **29**(4), 446–454 (2012)
8. Loukis, E., Charalabidis, Y., Androutsopoulou, A.: Promoting open innovation in the public sector through social media monitoring. Gov. Inf. Q. **34**(1), 99–109 (2017)
9. Fensel, D., Leiter, B., Stavrakantonakis, I.: Social media monitoring. https://oc.sti2.at/sites/default/files/SMM%20Handouts.pdf (2012). Accessed 12 Sep 2018
10. Janssen, M., Konopnicki, D., Snowdon, J.L., Ojo, A.: Driving public sector innovation using big and open linked data (BOLD). Inf. Syst. Front. **19**(2), 189–195 (2017)
11. LoRa-alliance: https://lora-alliance.org/ (2018). Accessed 12 Sep 2018
12. Jetter, H-C., Gallacher, S., Kalnikaite, V., Rogers Y.: Suspicious boxes and friendly aliens: exploring the physical design of urban sensing technology. In: Proceedings of the First International Conference on IoT in Urban Space (URB-IOT '14). Brussels, Belgium, pp. 68–73 (2014)
13. Nissenbaum, H.: Respecting context to protect privacy: why meaning matters. Sci. Eng. Ethics. **24**, 831–852 (2018)
14. Baldini, G., Botterman, M., Neisse, R., Tallacchini, M.: Ethical design in the internet of things. Sci. Eng. Ethics. **24**, 905–925 (2018)

15. Salim, F., Haque, U.: Urban computing in the wild: a survey on large scale participation and citizen engagement with ubiquitous computing, cyber physical systems, and Internet of Things. Int. J. Hum. Comput. Stud. **81**, 31–48 (2015)
16. Checkland, P.: System Thinking, System Practice. Wiley, Avon (1981)
17. Mikusz, M., Houben, S., Davies, N., Moessner, K., Langheinrich, M.: Raising awareness of IoT sensor deployment. In: Living in the Internet of Things: Cybersecurity of the IoT. Institution of Engineering and Technology, London (2018)
18. Weiser, M., Brown, J-S.: The coming age of calm technology. http://quicksilver.be.washington.edu/courses/arch498cre/2.Readings/2.Theory/CalmTech(Weiser%20&%20Brown,%201996).pdf (1996). Accessed 12 Sep 2018
19. Laurel, B.: Computers as Theatre, Second edn. Pearson Education, Crawfordsville (2014)
20. Gibson, J.J.: The Ecological Approach to Visual Perception. Houghton Mifflin, Boston (1979)
21. Norman, D.: The Design of Everyday Things. Basic Books, New York (1999)
22. Kaptelinin, V., Hedestig, U.: Breakdowns, affordances and indirect design: a study of videoconference learning environment in undergraduate education. In: Dirckinck-Holfeld, L., et al. (eds.) Analysing Networked Learning Practices in Higher Education and Continuing Professional Development, pp. 223–240. Sense Publisher, Rotterdam (2009)
23. Gaver, W.: Technology affordances. In: CHI'91 Conference Proceedings, pp. 79–84. ACM Press, New Orleans (1991)

Exploring New Digital Affordances of City Life

Rui José and Paula Trigueiros

1 Introduction

Cities are growing faster than ever before. As described by Townsend, they accelerate time by compressing space, letting us do more with less of both [1]. In general, they offer better infrastructures, but more importantly, cities are exciting because they offer options: where to go, what to do, what to see [2]. Nowadays, digital services play an increasingly central role in shaping the ways in which we collectively experience cities. Cities are labelled as 'digital', 'smart' or 'connected' to refer to the speed, quality, coverage or sophistication of those digital services and the new opportunities they create. These visions of the digital city are often connoted with major investments in enabling infrastructures, such as basic communication networks, urban sensors and many types of connected things. However, a 'smart city' is also a city that makes its citizens smarter by empowering them to reflect and act on their environment [3]. When considering urban infrastructures, we should thus also consider their role as enablers for open, spontaneous and creative uses or urban space. This can be critical for the ability of cities to engage and promote positive interactions with their citizens.

This stronger focus on people, as users, citizens or visitors, has been thoroughly explored by the fields or urban media and creative cities. For example, the Playable Cities initiative [4] has explored many and very sophisticated citizen engagement opportunities enabled by information technologies in urban space. However, despite

R. José (✉)
Algoritmi Research Centre, University of Minho, Braga, Portugal
e-mail: rui@dsi.uminho.pt

P. Trigueiros
Lab2PT, University of Minho, Braga, Portugal
e-mail: paula.trigueiros@arquitetura.uminho.pt

R. José et al. (eds.), *3rd EAI International Conference on IoT in Urban Space*,
EAI/Springer Innovations in Communication and Computing,
https://doi.org/10.1007/978-3-030-28925-6_5

being very exciting and engaging, these are normally transient installations, specifically designed for a very concrete context and, most of the times, for only a limited period of time. Ojala and Kukka claimed in 2009 that, despite the new possibilities of enriching the experience of the city through these new technologies, we still had to discover the 'killer' applications that would really bring the experience of the city to the next level [5]. Since then, researchers in different fields have explored numerous approaches to create engaging experiences in urban spaces. However, the general perception is that not much has changed or, at least, the major digital infrastructures that were envisioned 10 years ago are yet to produce visible and sustainable contributions to city life.

In this work, we want to abstract away from any visions of the city of the future and focus instead on the current reality of urban life. In particular, we are concerned with the infrastructural elements which are emerging today in many cities to enable a deeper connection with their inhabitants, even if they do not always correspond to the high-profile visions normally associated with smart cities. Our research goal is to uncover signs of these new forms of urban digital engagement that are actually making their way into the city fabric and impacting the behaviour of citizens and visitors. We will borrow the concept of Digital Affordance, as introduced to the HCI community by Donald Norman [6]. The Digital Affordances of a city are thus the 'visible cues' representing the digital layer of the city in physical space. They can be perceived and experienced directly by people and they provide particular engagement opportunities in urban space. To be aligned with the concept of affordance, their interactive experience must represent a common expectation for most people, at least in developed countries. They need to demonstrate, through their widespread presence and usage, that they are long-lived, economically sustainable, widely deployed and open to common use by people.

We aim to identify the most representative examples of this type of Urban Digital Affordances, analyse their key properties and synthesise the key concepts involved. By characterising these concepts, we expect to inform the design of new digital infrastructural elements that blend strong digital engagement with low-cost and large-scale deployments. To pursue this goal, we have taken the perspective of the city as viewed by its citizens and its visitors. We believe this to be the viewpoint that better represents the way the city affordances are actually available to be experienced. We have collected 93 photos of relevant digital elements across 4 European cities and analysed them according to their properties. The result is a taxonomy of 5 key concepts, which may contribute to guidelines for the design of new concepts for urban engagement.

2 Related Work

The concept of smart cities is often connoted with major digital infrastructures, leading many cities to embed new digital technologies into municipal infrastructures. From this perspective, cities can become smarter because they are equipped with embedded sensors, processors, actuators and smart 'things'. These are designed to sense and interact with the physical world (including the human users) and support real-time, guaranteed performance in safety-critical applications [7]. Cities instrumented with these digital monitoring infrastructures would be able to produce 'big data' and enable real-time analysis of city life to support more efficient, sustainable and transparent cities. In this work, we are not focused on technology itself. Our only concern is to explore how urban infrastructures, of any kind, can promote playful, low tech and spontaneous digital experiences.

Urban media is often proclaimed as a way to trigger, enhance and amplify urban experiences. It combines interdisciplinary skills to create entertaining and positive experiences in the urban space. The Playable Cities initiative is a representative example of this approach. Its many projects are conceived as new ways to put people and play at the heart of future cities around the world [4]. However, a common characteristic across most of those examples is the transient nature or very small scale of the respective deployments. These custom deployments have a strong focus on the citizen experience, but they often fail to consider scalability and economic viability.

Public displays might be seen as a relevant form of urban media that corresponds to long-term urban infrastructures. Foth et al. [8] claim they could serve as touch points between the city and its civic body, the citizens. Their unique capabilities to reach everyone around them, and without the restrictions imposed by personalisation algorithms, provide an opportunity to disseminate community and civic information. However, deploying and maintaining display networks is very costly. Most of the times, these displays are operated under advertising-driven business models and public-private partnerships, which do not leave much space for civic content or citizen creativity [9]. Koeman et al. [10] studied the use of chalk visualisations created on the pavement to represent community related information. This work shows that, when considering urban communication, it is valuable to consider affordances of various technologies, including the use of low-tech traditional information visualisations.

Appropriation of existing infrastructures for purposes other than the initially planned has been explored to promote scalability and easier service deployment. In some cases, this appropriation is achieved by merely augment it with a narrative for a virtual functionality. For example, the Hello Lamp Post [11] invites people to interact with existing urban furniture, e.g. lamp posts, bus shelters or parking meters, to express what they feel or want and talk about what happens in their neighbourhood. In other cases, appropriation involves the physical augmentation of an existing infrastructure. The Signpost Network [12] consists of a sensor network deployed as add-ons to existing signposts. One of the goals is to reduce the need for specific sensing infrastructures. The proposed devices can be attached to sign posts throughout the city and be powered by solar panels and battery storage. The Humble Lamppost toolkit for Action [13] is a relevant example of a modular approach to street furniture. It provides flexible deployment strategies that can more easily integrate different sub-sets of the urban infrastructure. This toolkit is conceived to help cities address deployment challenges by repurposing street lamppost as general purpose digital infrastructure. These are relevant examples of the importance of considering integration with existing infrastructures as a way to repurpose them and diminish the deployment impact of new services.

The DIY Media Architecture approach aims to place creative processes into the hands of laypeople [14]. The motivation is to use this as a path for more open and participatory approaches to engage communities. The authors suggest several approaches to help media architects to encourage the 'open-sourcing' of information. They also propose specific tools for allowing laypeople, not only to participate, but also to become active instigators of change in their own right.

Finally, cities can be also appropriated by digital technologies without any dependence on local infrastructures. Pokémon Go is an augmented reality location-based game. Its mega success has been a major landmark for the impact that unsolicited digital services can have in cities. Even though the app is meant to be about going outside and walking around, its design assumes that people will have the app opened all the time, and will be watching it even when they are running across the street to accomplish some goal of the game. This has been reported as an important safety hazard, involving accidents, crime and trespassing incidents, but it has also been described as the cause for a relatively rare large-scale shift in global human mobility patterns [15]. Using data from mobile phone networks, Graells-Garrido et al. [16] analysed the effect of Pokémon Go on the pulse of Santiago, in Chile. By comparing data collected before and after the game's release, they have shown how the city was affected in regard to: time spent outside; commuting patterns; or how 'the street' became a new place of leisure. The authors claim that these effects should be taken into account when considering long-term infrastructure investment or drafting public policies aimed at stimulating pedestrian traffic.

3 Research Methodology

To gain a deeper understanding about the digital infrastructure of todays' cities, we have conducted a field study where we sought to identify any types of Digital Affordances commonly available in urban spaces. The main criterion was to register any 'cues' at the urban landscape that could be directly associated with the digital world. These are Digital Affordances that would not be there if they had no role in digital services. As mentioned in the introduction, we were only concerned with elements that were available for the common citizen on a long-term basis and could thus be seen as being already embedded into current urban practices.

The initial part of this study was a photo survey, which took place across four European cities: Amsterdam, Munich, Braga and Porto. These cities are located in three different countries and they have different sizes, offering some diversity to the study. Field sessions in each city lasted between 1 and 3 days. In this process, we collected 93 photos, which, approximately, were evenly divided among those 4 cities.

We then conducted an annotation process, in which the two researchers created a textual description of each of those items and their usage context. The photos and these text annotations were then used as the main input to a qualitative analysis, in which we coded all the key concepts emerging from those images and their annotations. At first, the coding process was very open but, as new properties emerged and began to consolidate, we often went back to previous images to classify them according to those same codes. By the end of the process, a number of properties had emerged as general enough to be applied to all the photos, more specifically ownership, the type of connection between physical and virtual, situatedness in the urban space, and visibility in the physical and virtual worlds. We then aggregated the codes using an affinity diagram to identify major themes emerging from the data.

4 Results

In this study, we only considered hybrid services that combined a digital layer with some sort of physical presence in the city landscape. Still, even within this restricted scope, the results have clearly highlighted a sophisticated interplay between the physical and the digital layers of the city. We have found multiple and subtle combinations of these two dimensions, resulting in a variety of mixed concepts. In this section, we describe the 5 key concepts emerging from the analysis of the results, more specifically: Locative, Anchors, Hybrids, False Hybrids and Digital Counterparts.

4.1 Locative

The first set of digital infrastructures is related to Locative services. These are primarily digital services with a very lightweight connection to the city physical space. In most extreme cases, this connection is merely based on the virtualization of the physical space of the city and does not include any type of physical presence or representation of the service. Representative examples of this type of merely virtual connection include Google maps or Pokémon Go. Their affordances are deeply connected with the physical space of the city and they can have a major impact in how the cities are experienced by millions of people. However, since they do not require any physical presence to provide their affordances, they can grow much faster and quickly reach global scale.

Given their lack of a physical existence, we have not found any direct signs of their presence. However, we found indirect manifestations of their existence. Most of these Locatives services can greatly benefit from concrete connections with local entities, as they can substantially improve their own value proposition. For example, place seeding and crowdsourcing can be used to populate a database of

Fig. 1 Signs of Locative services across cities

local places, but much more value can be generated if local place owners are also actively involved.

These local partnerships allow those services to have a much more visible presence and also allow local place owners to get more value from the services. Therefore, place owners will often signal their association with these Locative services through stickers or other forms of merchandising. Even if these stickers do not offer a specific affordance themselves, they play an important role in creating awareness about the local availability of the Digital Affordances provided by those services.

Figure 1 shows multiple manifestation of Locative services. The image on the left shows a shop window announcing multiple Locative services associated with that venue. The image in the middle is from a restaurant and is announcing the availability of the Uber Eats food delivery service. The city centre tends to attract the most innovative services. The image on the right shows drivers of Uber Eats, who have chosen a downtown public walkway to strategically communicate and position themselves for quick response to incoming delivery tasks.

What all these services have in common is that they can have a huge impact in how the city is experienced, but they depend very little on any type of physical infrastructure or even city regulations. This trend will certainly raise new challenges for city planners, which will have to learn to deal in the best possible way with the positive and negative impacts that these new unplanned services will bring to city life.

4.2 Anchors

Anchors represent a type of digital affordance that aims to promote or restrict digital activity. Figure 2 shows some of the examples we found in our study, which correspond to embodied artefacts designed to trigger digital activity.

These examples have in common the presence of a physical artefact whose purpose is to trigger the use of digital services. The first two cases are clearly meant to promote social media presence through photo sharing. The street sign on the left

Fig. 2 Examples of artefacts designed to promote digital activity

Fig. 3 Anchors signalling possible restrictions

image is particularly explicit in its effort to drive people into taking ‘selfies’ at that spot. Big letters are slightly less explicit, but they are now commonplace at main landmarks, where they create the perfect setting for a ‘selfie’ that will promote the city brand through organic marketing and a physical hashtag. The third case was originally created to honour street sellers of newspapers. This type of human scale sculptures is meant to stimulate interaction with citizens, which can approach them physically. In the digital era, the iconic nature of this statue and its location near major tourist attractions have transformed it into a natural magnet for selfies. The final example creates awareness about the existence of a free Wi-Fi infrastructure in the city and its availability at that specific location.

These Anchors have in common that their existence is not relevant for the realisation of the respective affordances. People could still take photographs and share them, or use the local Wi-Fi, even if these artefacts were not present. However, their presence works as a major trigger for digital behaviour. Also, as simple triggers they can be open to many interpretations and digital services, enabling people to feel they are doing something unique and creative.

Anchors may also be used to signal areas of low-tech activity or special digital surveillance. These are areas where people may need or want to restrain some of their digital activity, as shown in Fig. 3.

In the left image, the goal is to create an area with a more simple and relaxed atmosphere, disconnected from the permanent distractions associated with the presence of mobile phones. In a sense, this is an anchor for disconnectedness. It may be interpreted as a sign of emerging practices, which in the future may lead to

Fig. 4 Hybrids infrastructures

digital free zones, very much like we now have pedestrian areas restricted to cars. The second image signals a surveillance area, giving people an opt-out option based on turning-off their phone or their Bluetooth/Wi-Fi connections. The final one is not providing any opt-out, but creates awareness about local surveillance and provides legal information about that process.

4.3 *Hybrids*

We have found a large number of physical artefacts that have a strong presence in urban landscapes, but are also strongly associated with digital services. These hybrid concepts manage to blend physical and digital elements in such an interdependent way that none of them would stand on its own. They are inherently associated with a specific physical affordance, but the digital component has become such a fundamental enabler that the essence of the service would not be possible without it. Figure 4 shows a few representative examples of these hybrids.

Charging stations for cars are inherently a physical affordance, but the need to find them, to know that they are available or even to reserve them, are absolutely fundamental for the experience of having autonomy when driving an electric car outside daily routes. A package delivery locker for on-line shoppers can significantly improve on-line sales of physical items by providing a delivery solution to people who may have difficulties in receiving them at a specific address. Local accommodation has become very popular worldwide and is having a huge impact in tourism trends. This is an obviously physical experience, but awareness, reservations and even physical access to the accommodation are increasingly digital. This affordance is clearly driven by global accommodation platforms, and such digital layer is an integral part of the whole concept. Likewise, most bike sharing systems require specific applications, which provide information on where to find available bicycles and how to unlock them and pay for them. While bike rentals may exist as traditional businesses, the concept of urban bike sharing is intrinsically connected to those Digital Affordances.

Fig. 5 Representative examples of False Hybrids

4.4 False Hybrids

A slightly different concept, which we called False Hybrids, is also a combination of a digital layer with the presence of physical artefacts on the streets. However, unlike real hybrids, these examples do not have an intrinsic dependence between digital and physical layers. In these False Hybrids, the physical element exists mainly for awareness and convenience, but it could be removed without affecting the essence of the service. Representative examples are shown in Fig. 5.

The kiosk on the left allows people to call a taxi. This is something that could easily be accomplished through many mobile apps or even through a simple phone call. Still, the kiosk creates awareness and convenience. The street kiosk in the second image can be seen as the modern equivalent of a phone booth. It also provides a communication point where people can access different types of information without having to use their own device. Interestingly, one of the most popular uses of these interactive displays is to take a photo and send it to an email address. While this can easily be achieved through the use of any common mobile phone, the specific affordance of doing it using a public infrastructure at a specific location seems to offer a popular value proposition. The live timetable service for public transportation is obviously very convenient, situationally valuable and reassuring. Still, all this information can easily be obtained by anyone with a connected mobile phone. Parking meters provide awareness and support key service functionality, such as payment and the emission of parking permits. However, they are increasingly being replaced or complemented with mobile applications that enable people to pay for parking time directly from their phones.

False Hybrids can be seen as a complement to purely digital experiences. Through their presence in physical space they can offer awareness about possibilities that could otherwise go unnoticed. Through the features provided by their physical instance, they can offer independence from mobile phones, network connectivity and to a larger extent from advanced digital literacy. For example, while there are many mobile applications that can be used to call taxis, not that many people will have them installed or even remember to use the mobile phone for that

Fig. 6 A specially designed False Hybrid

purpose, except maybe by calling a number. The physical element of False Hybrids provides an alternative for people who do not have a mobile device or who do not have one with the required networking or computational capabilities, who are not knowledgeable enough to use it for this purpose, or who have simply left their phone run out of battery. From this perspective, they can be seen as helping to reduce the risks of digital exclusion.

While some of these False Hybrids are likely to disappear as physical instances, many other may emerge to specifically explore their increased awareness and convenience. An interesting instantiation of a False Hybrid is deployed at the Van Gogh Museum in Amsterdam. This False Hybrid is not just some circumstantial step in technology migration. It has been specifically designed to create a place setting for an inherently Digital Affordance. At this museum, tickets are only available on-line and they need to be bought for specific visit times. Given these constraints, an occasional visitor may easily dismiss the possibility to visit the museum by considering that tickets are no longer likely to be available or that buying on-line tickets from the street might be too inconvenient. To overcome these barriers and grab these potential visitors, the museum created the space shown in Fig. 6.

The screens are showing the same ticket availability information available from the museum web site. Tickets still need to be bought on-line at the respective web site. However, free Wi-Fi is provided and guides people directly to that web page. To improve comfort, a shelter and some seats are also part of the design. This transforms this public space into a convenient ticket buying area, even without the presence of any ticket machines. It combines: opportunity awareness provided by the screens; an element of congregation, place making and convenience provided by proper urban furniture; and easy on-line access to the ticket office web page. Even

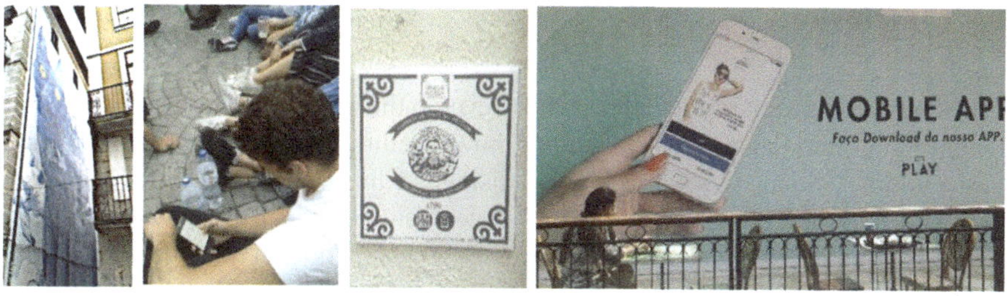

Fig. 7 Digital Counterparts

though the affordance of buying tickets is available anywhere, these local elements are all contributing to make this a special place for buying tickets.

4.5 Digital Counterparts

We also found many cases where the essence of the experience is inherently based on physical presence, but a Digital Counterpart can be used to add value to the overall experience. Typically, the value of this connection between the actual 'visible' locations and their virtual representations is either on the discoverability of the experience itself or in the possibility to complement the physical visit with a digital dimension. Figure 7 illustrates some very diverse examples of this concept.

Street art is an example of the type of urban experiences that are gaining popularity based on alternative digital city guides. The global visibility enabled by the digital element offers the unique capability to turn into mainstream what would otherwise only be known within small circles. The increased number of visitors to what before was a largely ignored element of the city, may often surprise residents and make them feel that they do not really understand what is going on in their own city. On the other hand, when the frequent presence of physical visitors becomes obvious, this can often lead to a renewed interest by the locals and the appropriation of the attraction for more conventional tourists.

Live events are also about being physically there, but they also combine a digital dimension that generates social media. They are not classified as Anchors because they are not primarily meant for that, but they can also generate a major digital presence. Small businesses increasingly try to have a web presence that they can push to visitors to enhance their visit and create a stronger bond with the venue. The smart tile in the image combines NFC and QR codes to drive people to a web page with the history of a local pastry. Also, many venues develop their own mobile app for visitors and incentivize people to install and use it as part of a richer visitor experience.

5 Conclusions

Cities want to be 'smart' and are eager to develop the infrastructures that may allow them to target that goal. These enabling infrastructures are often seen as providing basic Internet connectivity or generic sensing. In this work, we have shown that cities are also offering other types of infrastructures, which potentiate new and relevant Digital Affordances, even if many of them do not include any digital technology or any physical presence in the city space. More specifically, we have identified 5 core concepts of physical-digital services for urban spaces: Locative, Anchors, Hybrids, False Hybrids and Digital Counterparts. The relevance of the affordances offered by these 5 concepts has been validated by their widespread usage across the 4 cities in our study. Despite the diversity of their cultural and size characteristics, we have not observed major differences in the digital infrastructures of the 4 cities in our study. Most digital technologies seem to quickly find their way into very different urban contexts and cultural backgrounds.

Understanding the properties of these emerging infrastructures can help cities to be much more efficient in promoting creative uses of digital technologies in urban space. It may also help to develop new infrastructures that explore these same principles to accomplish new forms of citizen engagement. This should help to take citizen engagement to a systematic and common feature of city life, moving far beyond occasional high-profile installations or generic basic infrastructures.

References

1. Townsend, A.: Smart Cities: Big Data, Civic Hackers, and the Quest for a New Utopia. W. W. Norton & Company, New York (2013)
2. Zuckerman, E.: CHI keynote: desperately seeking serendipity. http://www.ethanzuckerman.com/blog/2011/05/12/chi-keynote-desperately-seeking-serendipity/ (2011). Accessed 20 July 2018
3. EIP SCC's Citizen Focus Action Cluster.: Inclusive smart cities: a European manifesto on citizen engagement (2017). https://eu-smartcities.eu/clusters/3/description Last accessed 2018
4. Nijholt, A.: Towards playful and playable cities. In: Playable Cities. Springer, Singapore (2017). https://doi.org/10.1007/978-981-10-1962-3
5. Ojala, T., Kukka, H.: A digital city needs open pervasive computing infrastructure. In: Workshop on Digital Cities 6 Concepts Methods and Systems of Urban Informatics at the 4th International Conference on Communities and Technologies CT 2009 (2009)
6. Norman, D.A.: The Psychology of Everyday Things. Basic Books, New York (1988)
7. Sun, Y., Song, H., Jara, A., Bie, R.: Big data analytics on smart and connected communities using internet of things. Int. J. Pharm. Technol. **8**(4), 19590–19601 (2016)
8. Foth, M., Tomitsch, M., Forlano, L., Haeusler, M.H., Satchell, C.: Citizens breaking out of filter bubbles. In: Proceedings of the 5th ACM International Symposium on Pervasive Displays - PerDis '16, pp. 140–147 (2016). https://doi.org/10.1145/2914920.2915010
9. Iveson, K.: Branded cities: outdoor advertising, urban governance, and the outdoor media landscape. Antipode. **44**(1), 151–174 (2012). https://doi.org/10.1111/j.1467-8330.2011.00849.x

10. Koeman, L., Kalnikaitė, V., Rogers, Y., Bird, J.: What chalk and tape can tell us: lessons learnt for next generation urban displays. In: Proceedings of the 3th International Symposium on Pervasive Displays (PerDis '14), p. 130 (2014). https://doi.org/10.1145/2611009.2611018
11. Nijholt, A.: Designing humor for playable cities. In: Procedia Manufacturing (AHFE), vol. 3, pp. 2175–2182 (2015). https://doi.org/10.1016/j.promfg.2015.07.358
12. Adkins, J., Campbell, B., Arbor, A., Ghena, B., Dutta, P.: The signpost platform for city-scale sensing. In: SenSys '17 Proceedings of the 15th ACM Conference on Embedded Network Sensor Systems, pp. 1–2 (2017). https://doi.org/10.1145/3131672.3136990
13. EIP SCC.: The Humble Lamppost Toolkit for Action. EIP SCC (2015). https://eu-smartcities.eu/initiatives/78/description. Last accessed Jul 2018
14. Caldwell, G.A., Foth, M.: DIY media architecture: open and participatory approaches to community engagement. In: Proceedings of the 2014 Media Architecture Biennale, pp. 1–10 (2014). https://doi.org/10.1145/2682884.2682893
15. Colley, A., Wenig, N., Wenig, D., Hecht, B., Schöning, J., Thebault-Spieker, J., et al.: The geography of Pokémon GO. In: Proceedings of the 2017 CHI Conference on Human Factors in Computing Systems - CHI '17, pp. 1179–1192 (2017). https://doi.org/10.1145/3025453.3025495
16. Graells-Garrido, E., Ferres, L., Caro, D., Bravo, L.: The effect of Pokémon Go on the pulse of the city: a natural experiment. EPJ Data Sci. **6**(1), 23 (2017). https://doi.org/10.1140/epjds/s13688-017-0119-3

A Comprehensive View on Quality Characteristics of the IoT Solutions

Miroslav Bures, Xavier Bellekens, Karel Frajtak, and Bestoun S. Ahmed

1 Introduction

To measure the quality of a System Under Test (SUT), various quality characteristics are being employed as standard industry practice, for instance, Jung et al. [16], van der Aalst et al. [27], Gucegliogiu and Demirors [11]. These characteristics are covering various quality aspects of the SUT, spanning from the presence of defects in SUT functionality to broader issues, ranging from usability and maintainability to the testability of the systems [27].

The importance of these characteristics lays in several functions:

1. They serve as managerial tools to measure the quality of SUT and contribute to making the quality assurance process more effective;
2. They emphasize different quality aspects besides the simple presence of software defects. Hence, they lead system engineers to focus on aspects like testability, maintainability, scalability, or others, which are not directly quantified by the presence of defects explicitly reported by the testing teams. However, these can have a significant impact on project or product success or failure;

M. Bures (✉)
FEE, Czech Technical University in Prague, Prague, Czech Republic

School of Design and Informatics, Abertay University, Dundee, UK
e-mail: miroslav.bures@fel.cvut.cz

X. Bellekens
School of Design and Informatics, Abertay University, Dundee, UK
e-mail: x.bellekens@abertay.ac.uk

K. Frajtak · B. S. Ahmed
FEE, Czech Technical University in Prague, Prague, Czech Republic
e-mail: frajtak@fel.cvut.cz; albeybes@fel.cvut.cz

R. José et al. (eds.), *3rd EAI International Conference on IoT in Urban Space*,
EAI/Springer Innovations in Communication and Computing,
https://doi.org/10.1007/978-3-030-28925-6_6

3. They help setting up an efficient test strategy for particular SUT, but most importantly they help managerial decisions regarding the quality aspects that are important and inform on which techniques shall be employed to prove SUT quality;
4. Considering the fact that the security and privacy are discussed as the main challenges of the current IoT solutions [8, 18], proper quality characteristics may help reducing cyber-security and privacy risk by revealing flaws and reducing the attack surface by ensuring that the correct operations are executed.

In the software testing and quality assurance industry, several sets of quality characteristics have been established and used. As an example, we can give *ISO/IEC 9126*, later replaced by *ISO/IEC 25010:2011* [16] or *TMap Next* [27]. As IoT systems differ from web-based software enterprise systems in a number of points and also bring several challenges specific to IoT infrastructures [8, 18], it is, therefore, relevant to revise these quality characteristics and quality metrics for IoT systems to capture their specificities. A first attempt has already been made [1–3, 9, 10, 12, 14, 15, 19, 21, 25, 26, 28]; however, these classifications focus specifically on heterogenic areas of IoT systems, applications, and viewpoint on the system quality. Hence, a more consolidated system is required, discussing the particulars of IoT domains and their intrinsic nature.

Generally, we need to distinguish between quality characteristics and quality metrics. **Quality characteristic** is a general category, describing a particular viewpoint on the SUT, which can be used in the test planning, test strategy, or test reporting. It is not defined by a particular formula which employs particular elements and quantities from the SUT model or facts from the test management process. Differently, **quality metric** is usually expressed as a formula, in which various facts from the testing process or SUT model is used (e.g., number of executed test cases, number of found defects, number of requirements covered by test cases, measured times in the tests, and others). Also, elements of SUT models used for test design purposes can be used (e.g., various metrics capturing structural test coverage).

In this paper, we attempt providing a unified classification of quality characteristics specifically dedicated to IoT architecture, ranging from the availability to the cyber-security and usability of the systems.

The remainder of this paper is organized as follows; Sect. 2 broadly introduces work on quality characteristics and metrics, while in Sect. 3 we provide a unified classification of quality characteristics focusing on IoT systems. Section 4 records selected points related to the consolidation process. The last Section concludes the paper.

2 Related Work

Currently, a set of individual attempts to categorize quality characteristics for IoT systems can be found, most of them focusing on a specific area or aspect of IoT systems, or not sufficiently focusing on the physical level of an IoT system. In this chapter, we discuss these works.

The TMap Suite (previously TMAP Next) is the body of knowledge for professional testers, created by Sogeti corporation, summarized quite an exhausting list of quality characteristics with the selection of these characteristics relevant to IoT testing [26]. However, as the major background of the company is in software testing, this list does not reflect some important networking and physical layer aspects of IoT systems. We can find these characteristics covered by other authors.

A quality model targeting cloud service called Cloudqual was defined by Zheng et al. [28]. This model is used to represent, measure, and compare the quality of the cloud service providers. It contains six quality dimensions in total—one subjective, i.e., usability, and the others objective—availability, reliability, responsiveness, security, and elasticity. Empirical case studies on three storage clouds were conducted. Similarly, the trust of cloud service providers is calculated using the proposed novel trust model based on past credentials and present capabilities of a cloud resource provider by Manuel [21]. Quality characteristics discussed in these studies related to the IoT cloud services can be used in the classification of these characteristics for general IoT systems.

Data quality metrics in pervasive environments were defined by Li et al. [19] and applied on real-world data sources to demonstrate the feasibility of the metrics. Previous data quality characteristics in database applications were not applicable to pervasive environments and the metrics proposed in QoC research were either unobservable or unadaptable to application requirements. Three metrics were redefined for pervasive environments, namely currency, availability, and validity, to quantitatively observe the quality of real-time data and data sources.

Regarding the security area, which is being frequently discussed as one of the most significant IoT challenges [8, 18], individual studies can be found. As an example, a study by Islam and Falcarin [15] can be given. The authors identified security requirements through asset-based risk management process to describe the security goal. Security of the IoT platform is one of the most important requirements, and the results of this study are applicable here.

Various security metrics are used to quantify the degree of freedom from a possibility of suffering damage or loss from a malicious attack. These key metrics have been defined by Abbadi [1]. Security and usability assessment of several authentication technologies are analyzed and summarized by Solie [25].

The applications of IoT bring new possibilities of what the user can achieve and experience. A subjective user's satisfaction with the application—quality of experience, QoE—will become new quality metrics the operators will have to consider. Ikeda et al. propose a framework of scalable QoE modeling for explosively increasing applications [14]. They defined two sets of metrics—physical metrics emerging in the IoT architecture and metaphysical metrics demanded by users.

The quality of the data at the device and network level is also covered by the literature. Banerjee and Sheth explore challenges in interpreting and evaluating the quality of data at informational and application levels [3]. Authors propose solutions at the different OSI layers to understand the factors affecting the quality of the system.

Cloud applications can scale up and down on demand depending on the application load, which is referred to as elasticity. This characteristic is discussed in the study by Han et al. [12]. In this study, elastic scaling approach making use of cost-aware criteria to detect and analyze the bottlenecks of the cloud-based applications along with adaptive scaling algorithm for cost reduction was presented.

The nature of the IoT platform where devices (especially mobile devices) are dynamically joining or leaving the network creates new issues in enforcing QoS of such platform. Gomes et al. discuss this scalability characteristic and propose a new approach for resource allocation focusing on the performance of the system when participants disconnect [10].

Another relevant characteristic, an information flow efficiency is explored in supply chain management by Badenhorst et al. [2]. A conceptual framework of indicators and data-oriented metrics to evaluate the efficiency of information flows in supply chains are introduced in this study.

Also testability of an IoT system, especially testability by automated tests shall be considered as a quality characteristic. Previous attempt to define metrics for automated testability has been done for web applications [5, 6], relevant for automated tests using the web-based user interface of the SUT. As IoT systems provide web-based user interfaces in many cases, this proposal can be applicable also to IoT context.

As the individual works discussed in this section focus rather on the IoT specific areas, on certain segments of the whole IoT platform, or does not reflect the quality aspects of IoT system in their full spectrum, a consolidated view has to be created to cover the whole spectrum of the IoT quality characteristics.

3 Proposed Classification

In the proposed classification, we merged several relevant proposals [1–3, 9, 10, 12, 14, 15, 19, 21, 25, 26, 28] into one unified view, which we enriched by several own suggestions of quality characteristics relevant for the IoT systems.

In the proposal, we followed several design principles:

(I) We added a physical device layer aspect to the classification, as this aspect becomes especially relevant in case of the IoT systems.

(II) We focused in special detail on Security, Privacy, and Usability aspects, as these areas are considered as being critical for the IoT domain [8, 18].

(III) We tried to minimize possible overlaps and duplications in the final proposed classification.

In this paper, we deliberately focus on quality characteristics instead of more detailed quality metrics. The reasons are the following: (1) the quality metrics might be too specific considering particular subdomain of IoT systems so that generalization might be not possible, and (2) much more SUT modeling information shall be required, making such attempt being out of the scope of the conference paper. Hence, in our consolidation, we abstracted some of the quality metrics from a subset of surveyed work (for instance [1, 14]) to the quality characteristic without biasing the original purpose and meaning of the metric.

Table 1 presents this consolidated view. Regarding the level of granularity, we decided to use two levels: main level quality characteristic (in Table 1 by **bold**) and second-level quality characteristic, being a subcategory of the main level (in Table 1 by *italics, indented*).

For several quality characteristics, synonyms have been used in the investigated literature. Also, due to our experience, several synonyms are used in the industry praxis. We put these synonyms to the footnote with a citation to the source paper (or a comment that the synonym is our suggestion based on the industrial praxis). The last column of the Table 1 presents the origin of the suggested quality characteristic. Word *own* in this column indicates that the quality characteristic is our suggestion based on the industrial experience and quality characteristics defined in test management methodologies for traditional software systems.

4 Discussion

In this section, discussed are several issues related to the approach taken and related questions.

Regarding the selection of the resources, to compile the presented consolidated view, we preferred works which are also consolidating the previous ideas, for instance, summary by Li et al. [19], which aggregates a set of previous works as [4, 17, 22–24].

During the creation of the presented consolidated view on IoT systems Quality Characteristics, interpretation of particular items may be different by individual authors. As an example, *Availability* can be discussed: this characteristic is described as "Uptime percentage of cloud services during a time interval" by Zheng et al. [28], or as a "Availability of the data sources, measured by a ratio of the number of attributes available to the total number of attributes" by Li et al. [19]. In such cases, we consolidated the metric to more general one, as, in the example of *Availability* was "Availability of the provided service or particular data supplied as the part of the service." Another example of this generalization is the *Data quality*, where Li et al. [19] understands this characteristic to cover all layers of the SUT spanning from physical layer to the user interface layer, whereas Ikeda et al. [14]

Table 1 Proposed unified classification of IoT systems quality characteristics

Quality characteristic	Description	Source
Accessibility	The extent to which the system can be handled by users with specific needs	Own
Availability	Availability of the provided service or particular data supplied as the part of the service	[19, 28]
Device lifespan expectancy	What is the estimated lifetime of a single HW device?	Own
Elasticity	The ability of the system to provide particular service on demand during a time interval	[28]
Installability	What is the estimated lifetime of a single HW device?	[26]
Ease of deployment	Effectiveness and efficiency with which the application can be deployed to devices of the system	Own
Interoperability[a]	Capability of the system, product, or device to interact with another system, product or device or to interchange data with it	Own, [26]
Maintainability	Effort needed to perform various maintenance tasks of the deployed system	[26]
Replaceability	Effectiveness and efficiency with which an invalid unit or device of the system can be replaced	Own
Updateability	Effectiveness and efficiency with which a unit or device be updated to the latest version	Own
Performance	The extent to which the system is able to handle a certain amount of data and concurrent user/device traffic	Own
Privacy	The extent to which the system maintains access to the user data corresponding to defined access rights by all involved parties and the extent to which the system prevents abuse of the user data	[15, 26]
Randomness	The extent to which the cryptographic algorithm used for protection generated random numbers (i.e., entropy size, or the randomness of the seed)	[15]
Data privacy[b]	The extent to which the data is safely stored with appropriate measures (i.e., Encyrption). This can also include the location of storage	[15, 26]
Data transmission	The extent to which the data is vulnerable to a replay or a Man in the Middle (MITM) attack. The extent to which the data being transmitted is encrypted with an appropriate algorithm	[15]
Access control	The extent to which the user has access to data, and the data he and others can modify	[15]

(continued)

Table 1 (continued)

Quality characteristic	Description	Source
Non-repudiation	The extent to which the system can guarantee that the data has not been modified	[15]
Proof of transaction	The extent to which a transaction can be proven to be from a user and the extent to which this user can be identified by other users if the data is leaked	[15]
Reliability[c]	The extent, in which the system is free from hardware and software defects, or other defects, which can lead to system failures	[26, 28]
Data quality[d]	Is the quality of the data provided by the system on the various levels (device, network, computing, and user interface) sufficient to enable correct run of the service?	[3, 14, 19]
Functional correctness	What is the error rate of the system in the sense of functional defects affecting the system processes and the procedures handling the data stored in the system?	Own
Up-to-dateness[e]	Are the data obtained from the system or device actual enough to enable correct operation of the service?	[19]
Trustworthiness	The extent to which the data provided by the system are trusted by its users	[19, 21]
Resource utilization[f]	The extent to which the resources required by the system were used in relation to the accuracy and completeness with which users of the system achieve their goals	[26]
Estimated energy efficiency	How long can the device operate without a power source? Does the device have a backup power source? Can it switch to passive mode when not needed?	Own
Responsiveness[g]	The extent to with the system handles a request within a required time interval	[28]
Satisfaction	The extent to which user needs are satisfied when the system is used in a particular operation	[26]
Scalability	The extent to which the system can adapt to new operational conditions, as the deployment model, size of processed data, user traffic conditions, added or removed devices, and others	[9, 10, 12]
Security	The extent to which the system protects its data so that any other party accessing to the system is given a level of access to this data, which is appropriate to the particular level of authorization	[1, 26, 28]
Attack surface	The number of interfaces provided to the user to access data and their associated security	[1]

(continued)

Table 1 (continued)

Quality characteristic	Description	Source
Given sense of control	The extent to which the user has access to control the device/application and the associated data collected and shared with the third party	[1]
Flaws vs time	The number of critical flaws found over a period during a review or after deployment	[1]
Data timeliness	The extent to which the data are backed up and can be retrieved by the users and/or forensic investigators	[2]
Data provenance	The extent to which the data is guaranteed to be provided by a trusted source	Own
Security compliance	The extent to which a system is compliant with a given security standard fit for its purpose (i.e., critical infrastructure, military, general public)	[1]
Testability	The extent to how easy is to design and conduct tests for the system, especially automated tests	Own
Usability	The extent to how easy, efficient, and enjoyable the interface of the system is to use and how efficiently the user interface contributes to support of the tasks user has to perform in the system	[25, 26, 28]
Subjective satisfaction	The extent to which a user is satisfied with both the software and its interface	[25]
Rate of user error	The extent to which the user encounter errors on the system or is required to perform an error action (i.e., reset a password, back-end error)	[25]
Speed to learn	The time required by the user to learn about the software and intrinsic characteristics. This can relate to the time for the user to perform an easy, medium, or hard action with a software	[25]
Support of secure behavior	The extent to which the security notifications and control are both enjoyable and understandable by a lay user. The user interface contributes to the cyber-situational awareness of the user	Own

[a] Also compatibility [26]
[b] Also confidentiality [26] or data store [15]
[c] Also correctness (our suggestion)
[d] Also precision [19] or data integrity [3]
[e] Also data validity (our suggestion)
[f] Also efficiency [26]
[g] Also time-behavior [26]

discuss this characteristic in context of the IoT devices. Similarly, we unified a concept of *scalability* understood differently by various authors [9, 10, 12].

In case of *Resource utilization* and *Efficiency* suggested by TMAP [26], these two categories seem rather overlapping, even if not equivalent. Hence, we decided to merge these both categories into final *Resource utilization*, as this characteristic express the idea better.

In another situation, we decided to add more specific quality characteristic, than was reported in works dealing with this topic previously. The example is *Accessibility*. In [26] and [28], *Accessibility* is implicitly understood as a part of *Usability*; however, according to the common understanding of these two concepts, for instance Henry et al. [13], our suggestion is to distinguish these two categories.

We made also generalization in case of *Confidentiality* suggested by TMAP [26]—we included this characteristic as a subtype of *Privacy* category, named as *Data Privacy*, as we merged this characteristic with *Data Store* suggested by Islam et al. [15].

A discussion can be made, if suggested *Up-time* subcategory of the *Reliability* does not duplicate *Availability*. As the *Up-time* related to particular IoT devices in the sense of their reliability, whereas *Availability* category describes the overall availability of the system, we decided to keep these two characteristics separated.

In the proposed categorization we decided to exclude characteristics metrics related to test coverage levels [27] as well as metrics for assessment of efficiency of test cases, for example [7, 20]. Such metrics might be discussed in the context of testing process efficiency; however, they do not directly relate to the quality characteristics of SUT (a relevant exception would be, when an automated test suite was considered as an SUT).

Another point can be raised regarding a question, if proposed two-level categories are appropriate to organize the discussed quality characteristics. In the current categorizations, only one level list is usually used, for instance, Jung et al. [16] or van der Aalst et al. [27]. However, specific focus on physical layer aspects, security, privacy, reliability, and maintainability of an IoT system led us to the identification of more relevant subcategories, which justify the proposed two-level structuring.

5 Conclusion

The usage of quality characteristics contributes to the better structuring of the testing process, helps in the test reporting and acts as a check-list for the test engineers aiding decision which quality aspects of the SUT to test. For these reasons, we consider it useful to provide a comprehensive view of the quality characteristics for the IoT systems, focusing on the specifics of these systems. As the previous work discusses rather individual areas of IoT systems and particular subareas of system quality, in this paper we provide a consolidated view. This effort involved extensive discussions arising from an attempt to consolidate the particular terminology used by various authors; we summarize this discussion in the Sect. 4. In the proposed classification we emphasized specific characteristics of the IoT system. Namely, we reflected physical device layer more intensely in comparison to standard software quality characteristics, for instance, Jung et al. [16], van der Aalst et al. [27] and we focused in special detail on Security, Privacy, and Usability aspects, as these areas are considered as being critical for the IoT domain [8, 18]. This focus makes the proposed IoT characteristics more relevant to IoT systems, compared to a case, when standard software quality characteristics would be used in testing of an IoT solution.

Acknowledgements This research is conducted as a part of the project TACR TH02010296 Quality Assurance System for Internet of Things Technology.

References

1. Abbadi, Z.: Security metrics what can we measure? In: Open Web Application Security Project (OWASP), Nova Chapter meeting presentation on security metrics, viewed, vol. 2 (2011)
2. Badenhorst-Weiss, J.A., Maurer, C., Brevis-Landsberg, T.: Developing measures for the evaluation of information flow efficiency in supply chains. J. Transp. Supply Chain Manag. **7**(1), 1–13 (2013)
3. Banerjee, T., Sheth, A.: IoT quality control for data and application needs. IEEE Intell. Syst. **32**(2), 68–73 (2017). https://doi.org/10.1109/MIS.2017.35
4. Bu, Y., Gu, T., Tao, X., Li, J., Chen, S., Lu, J.: Managing quality of context in pervasive computing. In: Sixth International Conference on Quality Software, QSIC 2006, pp. 193–200. IEEE, Piscataway (2006)
5. Bures, M.: Metrics for automated testability of web applications. In: Proceedings of the 16th International Conference on Computer Systems and Technologies, pp. 83–89. ACM, New York (2015)
6. Bures, M.: Model for evaluation and cost estimations of the automated testing architecture. In: New Contributions in Information Systems and Technologies, pp. 781–787. Springer, Berlin (2015)
7. Bures, M., Cerny, T., Klima, M.: Prioritized process test: more efficiency in testing of business processes and workflows. In: Kim, K., Joukov, N. (eds.) Information Science and Applications 2017, pp. 585–593. Springer Singapore, Singapore (2017)
8. Bures, M., Cerny, T., Ahmed, B.S.: Internet of things: current challenges in the quality assurance and testing methods. In: Kim, K.J., Baek, N. (eds.) Information Science and Applications 2018, pp. 625–634. Springer Singapore, Singapore (2019)
9. Buyya, R., Yeo, C.S., Venugopal, S., Broberg, J., Brandic, I.: Cloud computing and emerging it platforms: vision, hype, and reality for delivering computing as the 5th utility. Futur. Gener. Comput. Syst. **25**(6), 599–616 (2009)
10. Gomes, R., Bouloukakis, G., Costa, F., Georgantas, N., Da Rocha, R.: Qos-aware resource allocation for mobile iot pub/sub systems. In: International Conference on Internet of Things, pp. 70–87. Springer, Berlin (2018)
11. Guceglioglu, A.S., Demirors, O.: Using software quality characteristics to measure business process quality. In: International Conference on Business Process Management, pp. 374–379. Springer, Berlin (2005)
12. Han, R., Ghanem, M.M., Guo, L., Guo, Y., Osmond, M.: Enabling cost-aware and adaptive elasticity of multi-tier cloud applications. Futur. Gener. Comput. Syst. **32**, 82–98 (2014)

13. Henry, S.L., Abou-Zahra, S., Brewer, J.: The role of accessibility in a universal web. In: Proceedings of the 11th Web for All Conference, W4A '14, pp. 17:1–17:4. ACM, New York (2014). https://doi.org/10.1145/2596695.2596719
14. Ikeda, Y., Kouno, S., Shiozu, A., Noritake, K.: A framework of scalable QoE modeling for application explosion in the Internet of Things. In: 2016 IEEE 3rd World Forum on Internet of Things (WF-IoT), Dec 2016, pp. 425–429 (2016). https://doi.org/10.1109/WF-IoT.2016.7845393
15. Islam, S., Falcarin, P.: Measuring security requirements for software security. In: 2011 IEEE 10th International Conference on Cybernetic Intelligent Systems (CIS), Sept 2011, pp. 70–75 (2011). https://doi.org/10.1109/CIS.2011.6169137
16. Jung, H.W., Kim, S.G., Chung, C.S.: Measuring software product quality: a survey of ISO/IEC 9126. IEEE Softw. **5**, 88–92 (2004)
17. Kim, Y., Lee, K.: A quality measurement method of context information in ubiquitous environments. In: International Conference on Hybrid Information Technology, ICHIT'06, vol. 2, pp. 576–581. IEEE, Piscataway (2006)
18. Kiruthika, J., Khaddaj, S.: Software quality issues and challenges of Internet of Things. In: 2015 14th International Symposium on Distributed Computing and Applications for Business Engineering and Science (DCABES), pp. 176–179. IEEE, Piscataway (2015)
19. Li, F., Nastic, S., Dustdar, S.: Data quality observation in pervasive environments. In: 2012 IEEE 15th International Conference on Computational Science and Engineering, Dec 2012, pp. 602–609 (2012). https://doi.org/10.1109/ICCSE.2012.88
20. Li, N., Li, F., Offutt, J.: Better algorithms to minimize the cost of test paths. In: 2012 IEEE Fifth International Conference on Software Testing, Verification and Validation (ICST), pp. 280–289. IEEE, Piscataway (2012)
21. Manuel, P.: A trust model of cloud computing based on quality of service. Ann. Oper. Res. **233**(1), 281–292 (2015). https://doi.org/10.1007/s10479-013-1380-x
22. Manzoor, A., Truong, H.L., Dustdar, S.: On the evaluation of quality of context. In: European Conference on Smart Sensing and Context, pp. 140–153. Springer, Berlin (2008)
23. Neisse, R., Wegdam, M., Van Sinderen, M.: Trustworthiness and quality of context information. In: The 9th International Conference for Young Computer Scientists, ICYCS 2008, pp. 1925–1931. IEEE, Piscataway (2008)
24. Sheikh, K., Wegdam, M., Van Sinderen, M.: Middleware support for quality of context in pervasive context-aware systems. In: Fifth Annual IEEE International Conference on Pervasive Computing and Communications Workshops (PerComW'07), pp. 461-466. IEEE, Piscataway (2007)
25. Sollie, R.S.: Security and usability assessment of several authentication technologies. Master's thesis (2005)
26. TMap: a set of quality characteristics for iot testing. http://www.tmap.net/wiki/quality-characteristics#TitleLink18. Accessed 09 Aug 2019
27. van der Aalst, L., Roodenrijs, E., Vink, J., Baarda, R.: TMap NEXT: business driven test management. Uitgeverij kleine Uil (2013)
28. Zheng, X., Martin, P., Brohman, K., Xu, L.D.: Cloudqual: a quality model for cloud services. IEEE Trans. Ind. Inf. **10**(2), 1527–1536 (2014). https://doi.org/10.1109/TII.2014.2306329

Connected Vehicles in the Shark Tank: Rethinking the Challenge for Cybersecurity to Keep Pace with Digital Innovation

Nick G. Rambow, Steven Andorka, and Kira Rambow-Hoeschele

1 Introduction

With an increasing number of connectivity features in the car, it is transforming into a computer on wheels that is performing sophisticated data processing, containing over 100 million lines of code. Growing with every additional function, the software and hardware complexity will likely change the work requirements of Original Equipment Manufacturers (OEMs) in the future.

One differentiator from the information technology industry is beta testing, which is common business practice for information technology firms to identify weaknesses in software with the help of customers already using the product at an early stage. Whereas an error in the code or a hacking assault on a smartphone might not affect the user's life in a harsh manner, a malfunction of a connected car driving autonomously could lead to accidents and eventually to the death of people. Therefore, the topic of cyber security and system reliability in general is of high importance for OEMs [1–3].

N. G. Rambow
ESB Business School, Reutlingen, Germany

Portland State University, Portland, OR, USA

S. Andorka
Glasgow Caledonian University, School of Engineering, Glasgow, UK
e-mail: steven@andorka.de

K. Rambow-Hoeschele (✉)
Robert Bosch GmbH, Headquarters, Automotive Strategy, Stuttgart, Germany
e-mail: kira.rambow-hoeschele@de.bosch.com

R. José et al. (eds.), *3rd EAI International Conference on IoT in Urban Space*,
EAI/Springer Innovations in Communication and Computing,
https://doi.org/10.1007/978-3-030-28925-6_7

2 Definition of Security, Privacy, and Safety

As the amount and the sensitivity of data collected, transferred, and processed in the car will increase with a growing number of connectivity features, high security and reliability standards need to be ensured. To fully understand the threats and the solutions concerning car connectivity data, it is important to differentiate between three different areas: security, privacy, and safety. Security is related to the protection of technical information, preventing data manipulation or data theft. Privacy is concerned with the protection of private or personalized data. Safety is linked to the functional reliability of the car, ensuring that critical functions work under any circumstances.

Since the car is a safety-critical system, meaning that any vulnerability could lead to life-threatening situation, all three aspects, security, privacy, and safety, are affected by connecting the car [4–6].

3 Customer Needs and Supplier Requirements

According to studies, cyber security for connected cars is one of the most important issues for customers as they want their cars to be well protected and are sensitive toward threats that are linked to connectivity features. Moreover, 92% of Internet of Things (IoT) users are concerned about the topic of cyber security in general, while 54% of customers are afraid that hackers could manipulate the car. Thus, developing secure systems is not only important from a technical perspective but also to ensure that customers are willing to use those features [7, 8].

OEMs are facing the challenge that their products are becoming more and more complex due to a growing number of electronic control units (ECUs) and more lines of code coming with an increasing connectivity level. With many different parties involved along the supply chain, it is important to create certain security standards to ensure that no data lacks are created. Therefore, security needs to be considered from start of product development until market introduction and beyond when creating new products and features.

4 Shift in the Mobility Ecosystem and Value Chain

In the past, cars have been isolated systems but as they are increasingly becoming open and connected to a large ecosystem of many different players with a variety of access points, it is important to consider the challenges linked to these developments. Whereas it has been enough to protect cars from physical damage earlier, it is now also important to protect them against any kind of digital threats from the outside [9–12].

As OEMs are the ones who ultimately deal with the integration risk of features and bear responsibility, they need to ensure that security practices are consistently implemented throughout the whole value chain considering a variety of parameters.

Since they are facing the customers directly and will be blamed by the public if something goes wrong, it is of high importance for them to ensure best possible security standards. This can be achieved by establishing common industry standards through alliances ensuring a high level of security. However, it is not only the role of the manufacturers to develop high standards. Customers themselves also need to be aware of their responsibility and should always possess a certain level of cautiousness when using connectivity features, avoiding behaviors that can facilitate threats [13, 14].

5 Fundamental Work on Connected Vehicle Cybersecurity

5.1 Forms of Cyberthreats

When discussing the topic of cyber security, two different types of threats need to be considered: attacks by external parties such as hackers as well as system-based bugs that lead to malfunction. As cyberattacks on a car might not only lead to private data being stolen but also could make a difference between life and death, and this is an important topic that needs to be addressed. This statement was supported by the "Global Risk Report 2018" of the world economic forum [15] in which the issue of cyberattacks was ranked on the third place in terms of likelihood and number six in terms of impact.

Another identified risk was data theft, which was ranked at place four in terms of likelihood. There are a variety of cases where external parties could potentially attempt to get illegal access to car data, for example, pranksters, hacktivists, or vandals trying to demonstrate their skills or owners and operators programing software and collecting data without permission.

Furthermore, criminal organizations could strive to steal cars or cash to make transactions, to blackmail owners, or to encrypt or hijack cars while driving. However, not only criminals but also countries might have certain interests in car data for industrial espionage, surveillance, or economic and physical warfare [16–18].

Besides the threat of an external attack, the danger of system-based bugs needs to be recognized and measures need to be implemented to decrease the risk of malfunction to the least possible degree.

5.2 Conceptual Distinction Between Connectivity Features Related to Convenience and Driving Safety

At this point, it needs to be differentiated between convenience-related connectivity features and driving safety-related connectivity features such as "advanced driver assistance systems" (ADAS) that support the driver with autonomous driving functions.

Whereas bugs in convenience-related connectivity features could lead to harm in the form of data loss or wrong recommendations, driving safety-related features could lead to death of individuals in the worst case. Therefore, for every feature a certain level of reliability and security needs to be defined that it must fulfill before it can be released to the customer.

For instance, some ADAS need constant connectivity that has to be ensured to function properly at any time. Depending on the openness of the car's system and the degree of connectivity, the number and seriousness of security threats differ and need to be considered when thinking of security measures at the development stage [19–24].

5.3 Functionality of Connectivity Features

When breaking down the functionality of connectivity features of cars to basic terms, there are four parties involved along the data chain from the beginning until the end. First, sensors gather any kind of information depending on the feature. Second, gateways sort and filter the most important information. Third, data is sent and stored in a cloud platform database. Fourth, apps access data and use it to provide a certain service. When looking at the different sensors and interfaces of the car, there are a variety of ECUs and other surfaces that could be attacked [25, 26].

The ECUs and the controller area network protocol have been examples of vulnerable spots of the connected car in the past. As the controller area network protocol was a priority-based protocol where no authenticator field was needed, it was vulnerable to both spoofed packets and denial of service.

Moreover, obsolete proprietary encryption algorithms were used which used publicly known algorithms that were highly vulnerable to attacks using malicious code injection through firmware updates. In addition, while short-range wireless access via Bluetooth devices that allowed the execution of arbitrary code into the telematics system was used, long-range wireless access via the internet allowed hackers to take full control over vehicles.

6 Hacker's Perspective and Cybersecurity Solutions

According to research, the success of hacking relies on three fields: remote attack surfaces, cyber physical features, and in-vehicle network architecture. They need to be considered when designing devices linked to the car connectivity ecosystem, for example, the car itself, back-end solutions, smart devices, or the cloud [6, 27–30].

In the past, several OEMs had to recall their cars to improve the level of cyber security such as Fiat Chrysler that launched a recall of 1.4 million cars in the USA in 2015. Fortunately, they have recognized the importance of cyber security in cars and are working with other companies to analyze the security threats and to provide solutions.

Science suggests using technologies such as digital signatures, de-identification, and sophisticated encryption to foster security. Additionally, with the help of 5G networks, a higher reliability and security standard will be in place [31–34].

Smart and agile threat intelligence systems have been recommended as it is important to have not only energy reserves to maintain service in case of power failure but also network reserves that will continue working when being attacked [60]. To ensure that privacy is kept to the highest possible level, it has been suggested to minimize the amount of personal data that is stored as well as to be transparent about what is collected and how it is used. Furthermore, persons should have the possibility to securely delete any personal data or to set options upon what is being stored [35–37].

A variety of additional solution patterns are recommended by scientists to increase the level of security and privacy. First, by using air gaps, the points of entry can be reduced by isolating critical safety and control systems from non-essential systems. Such a clear separation can reduce the number of potential attack surfaces.

Second, through node encryption and lightweight cryptography it is possible to use cryptographic authentication and confidentiality services in internode communication to only grant access to data for certain parties making it difficult for external parties to access the system. Third, redundancy helps to stabilize the system and the failure of vital functions in case of hijacking for critical functions are implemented in more than one system.

Fourth, an analog backup can reduce the car's dependence on vulnerable electronic systems by serving as a mechanical last resort that is "non-hackable." Fifth, anomaly detection recognizes potential intrusions by tracking unexpected variations, relying on the predictability of data flows between certain systems.

Sixth, device and message authentication can help to identify when data is recorded, replayed, or spoofed by verifying the sender or receiver of information [38–40].

After having recognized a potential threat for the system, it will be possible to close security gaps through "over-the-air" updates in the future. These are highly important for ensuring a high security standard for connectivity features in the car as it shifts security measures from being static to being dynamic, enabling the system to react to changes in this volatile environment.

This is highly important since what has been considered secure one month ago might not be secure in the future. Through over-the-air update capability the time that is used to update a car is reduced immensely in comparison to today where customers potentially needing to visit a dealership to receive an update [21, 41].

7 Implications and Conclusion

With several technical solutions being available to solve the challenges of cyber security, the topic of privacy remains to be a very critical one for the public and needs to be closely evaluated. In the past, many critical voices toward privacy have been raised [42–44].

Therefore, when implementing new connectivity features in the car, one important question that needs to be answered for every individual product is the openness of it, meaning with how many interfaces it is connected and with whom it shares information. This is relevant not only from a functional perspective but also from an ethical one as every involved party has a certain responsibility in the car connectivity ecosystem [45].

References

1. Asaj, N., Schaub, F., Muter, M., Held, A., Weber, M.: ProTACD: a generic privacy process for vehicle development. In: 2013 12th IEEE International Conference on Trust, Security and Privacy in Computing and Communications (2013). https://doi.org/10.1109/TrustCom.2013.208
2. BBC.: Tesla and GM self-drive cars involved in road collisions. BBC. http://www.bbc.com/news/technology-42801772 (2018). Accessed 06 May 2018
3. Caroleo, B., Pautasso, E., Osella, M., Palumbo, E., Ferro, E.: Assessing the impacts of electric vehicles uptake: a system dynamics approach. In: 2017 IEEE 41st Annual Computer Software and Applications Conference (COMPSAC), pp. 790–795 (2017). https://doi.org/10.1109/COMPSAC.2017.67
4. IAV automotive engineering.: Zugriff erfolgreich abgewehrt. https://www.iav.com/automotion-magazine/automotion-03-2017/zugriff-erfolgreich-abgewehrt (2017). Accessed 06 May 2018
5. Joy, J., Gerla, M.: Internet of vehicles and autonomous connected car - privacy and security issues. In: 2017 26th International Conference on Computer Communication and Networks (ICCCN), pp. 1–9 (2017). https://doi.org/10.1109/ICCCN.2017.8038391
6. Koscher, K., Czeskis, A., Roesner, F., Patel, S., Kohno, T., Checkoway, S, et al.: Experimental security analysis of a modern automobile. In: 2010 IEEE Symposium on Security and Privacy, pp. 447–462 (2010). https://doi.org/10.1109/SP.2010.34
7. KPMG.: Security and the IoT ecosystem. https://assets.kpmg.com/content/dam/kpmg/pdf/2015/12/security-and-the-iot-ecosystem.pdf (2015). Accessed 06 May 2018
8. McKinsey & Company.: The Internet of Things: mapping the value beyond the hype. McKinsey Global Institute. https://www.mckinsey.de/files/unlocking_the_potential_of_the_internet_of_things_full_report.pdf (2015)
9. Atzori, L., Iera, A., Morabito, G.: The internet of things: a survey. Comput. Netw. **54**(15), 2787–2805 (2010). https://doi.org/10.1016/j.comnet.2010.05.010

10. Kneib, M., Huth, C.: On the fingerprinting of electronic control units using physical characteristics in controller area networks. In: INFORMATIK 2017, pp. 875–882 (2017). https://doi.org/10.18420/in2017_88
11. Lanotte, R., Merro, M.: A semantic theory of the Internet of Things. Inf. Comput. **259**(1), 72–101 (2018). https://doi.org/10.1016/j.ic.2018.01.001
12. Tabane, E., Zuva, T.: Is there a room for security and privacy in IoT? In: 2016 International Conference on Advances in Computing and Communication Engineering (ICACCE), pp. 260–264 (2016). https://doi.org/10.1109/ICACCE.2016.8073758
13. Ammar, M., Russello, G., Crispo, B.: Internet of Things: a survey on the security of IoT frameworks. J. Inf. Secur. Appl. **38**, 8–27 (2018). https://doi.org/10.1016/j.jisa.2017.11.002
14. Sicari, S., Rizzardia, A., Griecob, L., Coen-porisinia, A.: Security, privacy and trust in Internet of Things: the road ahead. Comput. Netw. **76**, 146 (2015)
15. World Economic Forum: The global risks report 2018, 13th edition. http://www3.weforum.org/docs/WEF_GRR18_Report.pdf (2018)
16. Dignan, P.: The future of the connected car. https://www.manufacturingglobal.com/technology/future-connected-car (2017). Accessed 06 May 2018
17. L'Heureux, A., Grolinger, K., Elyamany, H., Capretz, M.: Machine learning with big data: challenges and approaches. IEEE Access. **5**, 7776–7797 (2017). https://doi.org/10.1109/ACCESS.2017.2696365
18. Rossi, B.: How driverless cars can and will be hacked. http://www.information-age.com/how-driverless-cars-can-and-will-be-hacked-2-123460231/ (2015). Accessed 06 May 2018
19. Alaba, F., Othman, M., Hashem, I., Alotaibi, F.: Internet of Things security: a survey. J. Netw. Comput. Appl. **88**, 10–28 (2017). https://doi.org/10.1016/j.jnca.2017.04.002
20. Albishi, S., Soh, B., Ullah, A., Algarni, F.: Challenges and solutions for applications and technologies in the Internet of Things. Proc. Comput. Sci. **124**, 608–614 (2017). https://doi.org/10.1016/j.procs.2017.12.196
21. Khurram, M., Kumar, H., Chandak, A., Sarwade, V., Arora, N., Quach, T.: Enhancing connected car adoption: security and over the air update framework. In: 2016 IEEE 3rd World Forum on Internet of Things (WF-IoT), pp. 194–198 (2016). https://doi.org/10.1109/WF-IoT.2016.7845430
22. Luthra, S., Garg, D., Mangla, S., Berwal, Y.: Analyzing challenges to Internet of Things (IoT) adoption and diffusion: an Indian context. Proc. Comput. Sci. **125**, 733–739 (2018). https://doi.org/10.1016/j.procs.2017.12.094
23. Müller, D.-R.: Safety, security, privacy und continuity – Die Überraschung kommt zum Schluss. https://www.informatik-aktuell.de/betrieb/sicherheit/safety-security-privacy-und-continuity-sspc-die-ueberraschung-kommt-zum-schluss.html (2017). Accessed 06 May 2018
24. Sierra Wireless: Fueling connected cars from the automotive industry. https://www.sierrawireless.com/applications/automotive-and-transport/automotive/ (2018). Accessed 08 May 2018
25. Google: Designing a connected vehicle platform on cloud IoT core. https://cloud.google.com/solutions/designing-connected-vehicle-platform (2017). Accessed 10 May 2018
26. Intel: The state of security in the connected car. https://www.intel.de/content/www/de/de/automotive/automotive-security-best-practices-white-paper.html?_ga=2.219752217.1243940380.1526300302-362171776.1526300302 (2017)
27. Baldwin, R.: OwnStar car hacker can remotely unlock BMWs, Benz and Chrysler. https://www.engadget.com/2015/08/13/ownstar-hack/ (2015). Accessed 12 May 2018
28. Bécsi, T., Aradi, S., Gáspár, P.: Security issues and vulnerabilities in connected car systems. In: 2015 International Conference on Models and Technologies for Intelligent Transportation Systems (MT-ITS), pp. 477–482 (2015). https://doi.org/10.1109/MTITS.2015.7223297
29. Miller, C., Valasek, C.: A survey of remote automotive attack surfaces. http://illmatics.com/remote%20attack%20surfaces.pdf (2014). Accessed 11 May 2018

30. U.S. Department of Homeland Security: National Cybersecurity and Communications Integration Center/Industrial Control Systems Cyber Emergency Response Team review. https://ics-cert.us-cert.gov/sites/default/files/Annual_Reports/Year_in_Review_FY2015_Final_S508C.pdf (2015). Accessed 13 May 2018
31. Ericsson: 5G security. https://www.ericsson.com/assets/local/publications/white-papers/wp-5g-security.pdf (2017)
32. Girson, A.: IoT has a security problem - will 5G solve it? https://www.wirelessweek.com/article/2017/03/iot-has-security-problem-will-5g-solve-it (2017). Accessed 14 May 2018
33. Huawei: 5G security: forward thinking Huawei white paper. http://www.huawei.com/minisite/5g/img/5G_Security_Whitepaper_en.pdf (2015)
34. Li, S., Xu, L., Zhao, S.: 5G internet of things: a survey. J. Ind. Inf. Integr. **10**, 1 (2018). https://doi.org/10.1016/j.jii.2018.01.005
35. Jafarnejad, S., Codeca, L., Bronzi, W., Frank, R., Engel, T.: A car hacking experiment: when connectivity meets vulnerability. In: 2015 IEEE Globecom Workshops (GC Wkshps), pp. 1–6 (2015). https://doi.org/10.1109/GLOCOMW.2015.7413993
36. McAfee: Automotive security best practices. https://www.mcafee.com/de/resources/white-papers/wp-automotive-security.pdf (2016)
37. Yaqoob, I., Ahmed, E., Rehman, M., Ahmed, A., Al-garadi, M., Imran, M., Guizani, M.: The rise of ransomware and emerging security challenges in the Internet of Things. Comput. Netw. **129**(2), 444–458 (2017). https://doi.org/10.1016/j.comnet.2017.09.003
38. Studnia, I., Nicomette, V., Alata, E., Deswarte, Y., Kaaniche, M., Laarouchi, Y. (2013). Survey on security threats and protection mechanisms in embedded automotive networks. In: The 43rd Annual IEEE/IFIP International Conference on Dependable Systems and Networks (DSN-2013). https://doi.org/10.1109/DSNW.2013.6615528
39. Martinelli, F., Mercaldo, F., Nardone, V., Santone, A.: Car hacking identification through fuzzy logic algorithms. In: 2017 IEEE International Conference on Fuzzy Systems (FUZZ-IEEE), pp. 1–7. https://doi.org/10.1109/FUZZ-IEEE.2017.8015464
40. Förster, D.: Discussing different levels of privacy protection in vehicular ad-hoc networks. Ulmer Informatik-Berichte, pp. 31–32. https://pdfs.semanticscholar.org/974c/e32b617c11a18998fef3d543a8a35807a1ee.pdf#page=31 (2015)
41. Conti, M., Dehghantanha, A., Franke, K., Watson, S.: Internet of Things security and forensics: challenges and opportunities. Futur. Gener. Comput. Syst. 544–546 (2018). https://doi.org/10.1016/j.future.2017.07.060
42. Foy, H.: Connected cars raise privacy and safety worries. Financial Times. https://www.ft.com/content/e663c6fa-643b-11e4-bac8-00144feabdc0 (2014). Accessed 10 May 2018
43. Ramsey, M., Boudette, N.: Auto makers face questions over privacy in connected cars (W. S. Journal, Editor). https://www.wsj.com/articles/auto-makers-face-questions-over-privacy-in-connected-cars-1389890063 (2014). Accessed 04 May 2018
44. Wall, M.: Is your connected car spying on you? (BBC, Editor). http://www.bbc.com/news/business-29566764 (2014). Accessed 04 May 2018
45. Roman, R., Zhou, J., Lopez, J.: On the features and challenges of security and privacy in distributed internet of things. Comput. Netw. **57**(10), 2266–2279 (2013). https://doi.org/10.1016/j.comnet.2012.12

Part III
Methods and Technologies for IoT

AudioIO: Indoor Outdoor Detection on Smartphones via Active Sound Probing

Long Wang, Josef Roth, Till Riedel, Michael Beigl, and Junnan Yao

1 Introduction

In the past decade, we have seen the flourish of Internet of Things (IoT) and its applications in urban spaces. The increasing deployments of IoT technologies and the rise of the smart city are giving birth to more and more smart application cases [2, 7, 11, 14]. Context status is critical and fundamental information for ubiquitous and mobile computing systems and context-aware IoT applications [12, 29]. "Context" consists of a wide range of aspects such as location, time, surrounding environment, and so on. The rapid growth of smartphones is driving increasing interests in context-aware applications [16, 17, 21].

One of the most fundamental contextual information is whether the device is in indoor or outdoor environment. It makes a huge difference if you are standing in front of a shop or in a shop. Further, the availability and capabilities of different technologies vary strongly between these two environments. The knowledge about the Indoor Outdoor (IO) status enables the choice of appropriate technologies. This leads to benefits such as improved user experience and saving battery energy. For instance, the device can trigger reminder, change the working mode, switch between GPS based navigation and indoor navigation schemes when user enters an indoor environment or vice versa. Further, the device can save more energy by turning off GPS module in indoor environments such as a metro station. Existing IO detection approaches commonly make use of GPS signal, wireless signal, and other sensor data to determine IO status. The GPS based methods are typically not

L. Wang (✉) · J. Roth · T. Riedel · M. Beigl
Karlsruhe Institute of Technology, TECO, Karlsruhe, Germany
e-mail: wanglong@teco.edu; roth@teco.edu; riedel@teco.edu; beigl@teco.edu

J. Yao
Institute of System Engineering, Beijing, China

R. José et al. (eds.), *3rd EAI International Conference on IoT in Urban Space*,
EAI/Springer Innovations in Communication and Computing,
https://doi.org/10.1007/978-3-030-28925-6_8

energy efficient [8, 9, 19, 31]. The wireless signal based methods generally rely on existing infrastructures [4, 25, 32, 34]. The other sensor fusion schemes either partially rely on wireless infrastructures, or remain questionable when being applied in new environments [1, 13, 18, 33].

We propose AudioIO to address the IO detection problem for smartphones through active sound probing and machine learning. Active sound probing is a common technique in room acoustic to gain information about rooms [10]. We use the embedded speaker on smartphone to emit predefined sound probing signal (MLS and chirp signal) and record the reflected sound using the embedded microphone. The reverberation is determined by characteristics of the environment, like the size and shape of the space, the material of its surfaces and its interior. Intuitively indoor outdoor environments differ from each other regarding to volume and shape of the space. We conduct experiments to collect data in different locations. Audio features are extracted from the reverberation and feed to a SVM classifier with labels. We investigate the performance of AudioIO against noise level, device type, and probing signal.

To summarize, the contributions in this paper are:

1. We apply active sound probing in IO detection problem and provide comprehensive analysis. Although the idea of sound based IO detection was proposed in [27], it was roughly an idea and no in-depth analysis was provided. To the best of our knowledge, we are the first to comprehensively investigate the application of active sound probing technique in IO detection.
2. We implement an android APP for data collection and conduct experiments to collect data samples in various real daily scenarios. A data set consisting of more than 1000 labeled data samples is constructed.
3. We thoroughly evaluate the performance of SVM classifier on the data set with different noise levels, device types, and probing signals. Above 90% accuracy is achieved across different settings.

The rest of the paper is organized as follows: Sect. 2 presents related work. AudioIO is introduced in Sect. 3. The implementation is described in Sect. 4 and evaluation results are presented in Sect. 5. We conclude our work in Sect. 6.

2 Related Work

2.1 GPS Based Methods

GPS signal is highly dependent on the line-of-sight (LOS) paths between the device to GPS satellites. It is well known that the GPS signal is poor in indoor environments as the LOS paths of GPS signal are blocked. On the other hand, the LOS paths are free in most outdoor scenarios. According to these facts, the localization accuracy of GPS or the availability of GPS signal was exploited to determine if a device was in an indoor or outdoor environment [8, 9, 19, 31].

Despite that it is intuitive and easy to implement GPS based methods, they suffer from several disadvantages. Radu et al. identified the GPS chipset as the sensor with the highest power consumption among the evaluated sensors [22]. The battery capacity is still limited on state-of-art mobile phones and most users dislike applications which drain the battery. Secondly, the intuition behind these methods is not always reliable. For instance, the GPS signal is rather good if the device is in the indoor environment with large windows. Meanwhile, the GPS signal can be blocked by surrounding mountains if the device is in a valley. In these circumstances, GPS based methods may give misleading results. In the third place, it takes normally around 1 min to launch the GPS module. As a result, it does not fit for real-time applications.

In comparison, the acoustic interface on smartphone is generally more energy efficient compared with GPS module. It was reported in [5] that the power consumed by GPS module was 143.1–166.1 mW, meanwhile the power consumed by audio subsystem during audio playback with maximum volume was only 33.1 mW. Another evidence is that Nokia 95 smartphone can support telephone conversation for more than 10 h if battery is fully charged, but a turned-on GPS receiver can completely drain out the same battery within 6 h whether getting GPS readings or not [30]. Further, AudioIO enables real-time IO detection by analyzing echoes of emitted probing signal in the environment.

2.2 *Wireless Signals*

Shtar et al. [25] presented a method for continuous indoor outdoor environment detection on mobile devices based solely on WiFi fingerprints. The method assumes no prior knowledge of the environment. The model trained with the data collected for just a few hours on a single device was applicable for unknown locations and new devices. WifiBoost [4] made use of a machine learning meta-algorithm that combined an adequate ensemble of simple classifiers (so called weak learners) to improve the overall performance. An average error rate around 2.5% was achieved in the evaluation. However, the classifier needs to be created for each building and surrounding area through measurements and labeling each measurement point, especially in those cases where there is no previous fingerprinting database. Building such a database is not a trivial work.

Wang et al. [32] applied a machine learning algorithm to classify the signal strength of neighboring cellular base stations in different environments, and identified the current context by signal pattern recognition. 100% accuracy was reported in the results for identifying open outdoors, semi-outdoors, light indoors, and deep indoors.

In [34], low-power iBeacon technology was leveraged to develop an accurate, fast response, and energy-efficient scheme for indoor outdoor detection. The transitions between outdoor and indoor were achieved by comparing the Received Signal Strength of two pre-deployed Bluetooth beacons at two sides of each entrance.

Generally speaking IO detection methods based on different wireless interfaces rely on pre-deployed infrastructures, which limits their application scenarios. In comparison, AudioIO does not rely on any infrastructures.

2.3 Multiple Sensors

Since a single sensor might not be able to tackle all the application scenarios, data from multiple sensors such as accelerometer, proximity and light sensor, wireless receiver, and magnetometer were exploited for IO detection [18] [1, 13, 22, 33]. IODetector [33] combined data from three lightweight sensors (light sensor, cell tower signal strength, and magnetic sensor) to develop an extensible IO detection framework which did not require a training phase. Although acceptable error rates were achieved, Radu et al. [22] criticized IODetector for its hard-coded thresholds which might not work with new devices and new environments. Instead Radu et al. proposed a semi-supervised training method to improve IO detection accuracy across different devices and environments.

2.4 Other Methods

In [15], the embedded digital camera on mobile phone was utilized for IO detection. The developed gentle boosting classifier achieved an error rate of 1.7% for indoor and 10.8% for outdoor scenes. Beside, a feed forward neural network was trained with GIST feature of images to address the IO detection problem [28]. These methods help in generating semantic IO labels for images, but do not work for tracking and other real-time application cases.

Sung et al. [27] developed a sound based IO detection method using chirp signal. A simple classifier was developed on a static threshold. This work was rather simple and straightforward, no comprehensive analysis was provided.

3 AudioIO Architecture

We present the architecture of AudioIO in this section. The smartphone emits a short audio probing signal via its speaker and measures the reverberation of the environment. The reverberation is further processed as features of the environment. Finally a SVM model is trained on the features with IO labels. Figure 1 shows the architecture of AudioIO. It consists of reverberation measurement, preprocessing, and classification.

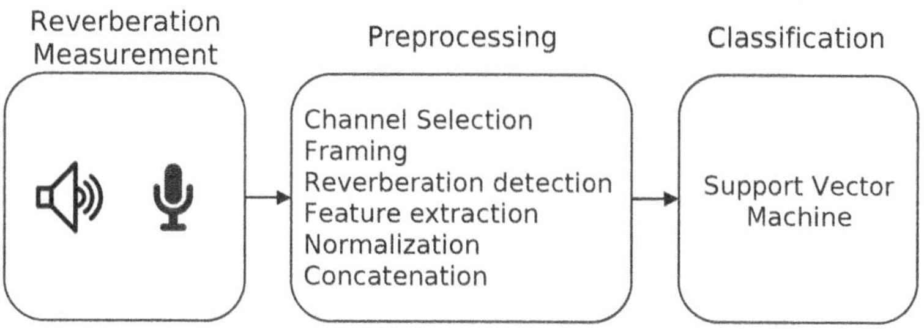

Fig. 1 AudioIO architecture

3.1 Active Sound Probing

Active sound probing is a common technique in room acoustic to gain information about rooms [10]. A short probing signal is played and the reflected sound is recorded. The reflections are determined by characteristics of the environment, like the size and shape of the space, the material of its surfaces and its interior.

A Dirac Impulse is theoretically the best probing signal, the impulse response defines the response of a linear time-invariant system for all frequencies [20]. However in practice it is impossible to generate and play a Dirac Impulse on smartphones. Therefore a range of other probing signals have been proposed. Existing literatures [10, 24] suggest that the Maximum Length Sequence (MLS) signal and the chirp signal are promising for Active Sound Probing. In this paper both signals are used for IO detection, the performance of both signals are compared in evaluation part.

Maximum Length Sequence (MLS) A Maximum Length Sequence is a type of pseudo-random binary sequences [26]. MLS is known for its robustness against background noise in the measured location. They are bit sequences generated using maximal linear feedback shift registers. The shift register can be represented by the following recursive function:

$$a_m[n+1] = \begin{cases} a_0[n] \oplus a_1[n] & m = 3 \\ a_{m+1}[n] & otherwise \end{cases}$$

where $\oplus$ denotes the XOR operation. MLS is characterized by its order M. The length of the MLS is $L = 2^M - 1$. In this paper, we chose the MLS signal with a common parameter set [24]. M is set as 15 and the sampling frequency is 44.1 kHz. Therefore the MLS signal is approximately 743 ms.

Chirp Signal A chirp is a signal in which the frequency increases (up-chirp) or decreases (down-chirp) with time. Studies have shown that the chirp signal shows a good performance in unoccupied rooms and it is easy to implement it [26]. In a linear chirp, the instantaneous frequency $f(t)$ varies exactly linearly with time:

$$f(t) = f_0 + kt \tag{1}$$

where f_0 is the starting frequency (at time $t = 0$), and ks is the rate of frequency change.

$$k = \frac{f_1 - f_0}{T} \tag{2}$$

where f_1 is the final frequency and f_0 is the starting frequency. T is the time it takes to sweep from f_0 to f_1.

We use chirp signal which ranges from 20 Hz to 20 kHz and lasts 100 ms, since Fan et al. [10] tested different sweep lengths and suggested this length as good balance between intrusiveness and capturing enough information.

3.2 Preprocessing

After the sound reflections are captured, the recording has to be processed for classification afterwards. The raw recording itself contains too much irrelevant information which would interfere with the classification. The goal of the preprocessing is therefore to extract relevant features for the IO detection.

Channel Selection Modern smartphones are often able to record audio signals in stereo. The relative position of the two microphones to the audio source and the orientation of the phone during measurement lead to a different quality of the recorded signal. As shown in Fig. 2, the amplitude of the signal in channel 2 is larger than that in channel 1 due to its proximity to the speaker. Intuitively the microphone closer to the speaker has similar orientation. As a result, it can receive more reflections of the probing signal. An energy detection algorithm is used to recognize the channel with more energy. The channel selection algorithm first calculates the FFT magnitudes for each channel. The magnitudes are then summed up for each channel. Finally the channel with the higher magnitude sum is chosen for further processing.

Framing For the sake of reverberation detection and feature extraction, the audio signal is split into smaller frames, which are then analyzed separately. The window length of each frame is set as 30 ms and a sliding window with 50% overlap is used to avoid information loss caused by signal smoothing. Each frame is smoothed with a Hamming filter.

Reverberation Detection Since it is not possible to synchronize the usage of speakers and microphones on Android mobile phones, it is not reasonable assume a fixed point in the recording as starting point for the reverberation. As shown in Fig. 3, the reverberations in two data samples arrived at different time which cannot be estimated beforehand. To detect the start and end of the reverberation, we first

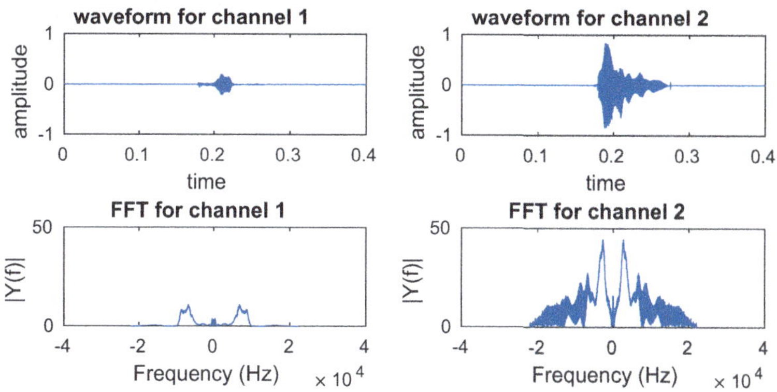

Fig. 2 Difference between data in two channels in one data sample

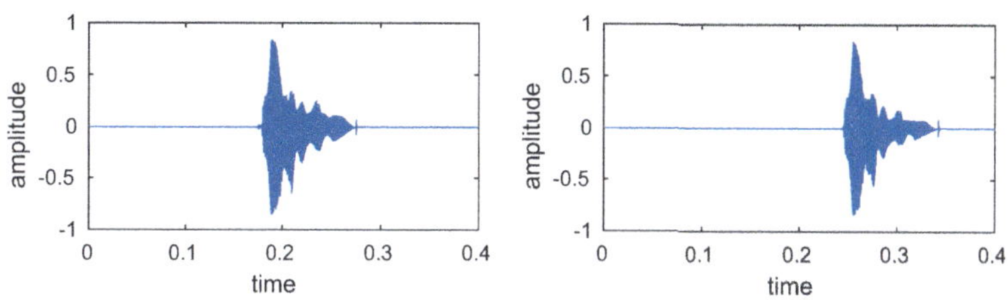

Fig. 3 Timing difference of reverberations in two data samples

calculate FFT for each frame and then sum the magnitudes of FFT. A predefined threshold is used to detect the start of the reverberation. Considering the sound attenuation in the space, we ignore the reflections from objects that are more than 50 m away from the smartphone. The length of the reverberation is set to the length of the probing signal plus additional 300 ms.

Feature Extraction Subsequently, audio features were extracted for each frame. We adopt Mel-Frequency Cepstral Coefficients (MFCC) with 12 coefficients as the feature for each frame. Mel-Frequency Cepstral Coefficients (MFCC) are commonly used features in room acoustic and audio classification [3, 10, 23, 24]. The data of an audio signal is reduced to a more compact form without losing too much information. MFCC represent only small, but most important parts of the frequency spectrum. In the next step the extracted features are normalized in each frame i by $F_i = \frac{f_i - m_i}{\sigma_i}$, where f_i is the raw feature vector, m_i and σ_i are the mean value and the standard derivation of the vector respectively. Finally, the normalized vectors from all the frames are concatenated to one feature vector $F = \{F_1, F_2, \ldots, F_n\}$.

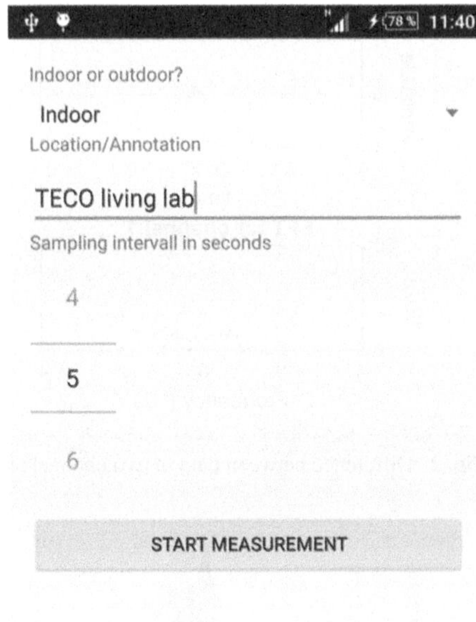

Fig. 4 Screenshot of the developed Android APP

3.3 *Classification*

The classification aims to generate a prediction of the current space of the smartphone if it is indoor or outdoor based on the extracted feature vector. Indoor outdoor detection falls into the category binary classification in machine learning. We use the Support Vector Machine (SVM) classifier for this task, the implementation of SVM is based on libSVM [6].

4 Experiment and Data Collection

4.1 *Experiment Setup*

The data for the evaluation is collected in a controlled experiment for which a dedicated Android APP is developed as shown in Fig. 4. This APP enables the mobile phone playing predefined MLS and chirp signal in different volume, measuring the response of the surrounding environment and labeling the data sample. The response is recorded with a sampling rate of 44,100 Hz, two channels, and a depth of 16 bit. To ensure the response is recorded completely, the recording is triggered before the probing signal is played. The measurements are saved as RIFF WAVE files to avoid information loss due to compression.

Measurement Locations The data is collected at different locations inside and around our campus field as shown in Fig. 5. The measurement locations are chosen

Fig. 5 Measurement locations. (**a**) Between buildings. (**b**) Canteen yard. (**c**) Woods. (**d**) Lab office. (**e**) Lab corridor. (**f**) Canteen hall

with the goal to cover different daily usage scenarios. In the indoor environments several room sizes were covered, ranging from offices and corridor in the lab building to the big dining hall in the canteen. The same criteria applies for the choosing of different outdoor locations. An example for a wide open space is the front yard of the canteen, whereas a road in woods is chosen to provide an example for an outdoor environment containing lots of objects. A measurement location between two near-standing buildings served as example environment for an outdoor scenario which could be confused with a closed room by the detection method.

Noise Generation The influence of noise is simulated using a generated white noise signal since white noise covers all the frequency band. The user holds the smartphone in one hand to emit and record audio signal, and holds a tablet in the other hand to play the white noise. Three different levels of noise are generated: quiet, medium, and loud. For the quiet level no white noise is generated. The volume of the generated white noise for the medium level is 40 dB. For the loud level the generated white noise has a volume of approximately 50 dB. The noise volume is measured using an APP on one of the recording devices in the same setup for the measurements.

Data Collection The data is recorded on three different devices: Google Nexus 7 (Nexus), RedMi Note (Mi), and Sony Xperia Z3 (Xperia). At each location a series of samples are recorded to cover the combinations of used devices and noise levels. The samples are taken while walking around through the location. We take the same route during each series as exactly as possible. A sample is recorded every 5 s while

Table 1 Statistics of collected data samples

		MLS			Chirp		
Device	Noise level	Indoor	Outdoor	Sum	Indoor	Outdoor	Sum
Mi	Loud	43	25	68	44	25	69
	Medium	36	25	61	36	25	61
	Quiet	37	24	61	37	24	61
Nexus	Loud	49	25	74	52	26	78
	Medium	45	25	70	47	25	72
	Quiet	44	20	64	45	21	66
Xperia	Loud	51	25	76	53	26	79
	Medium	48	27	75	48	30	78
	Quiet	50	19	69	50	19	69
Total		403	215	618	412	221	633

the probing signal is alternated between the chirp signal and the MLS signal. A detailed statistics of the data samples can be seen in Table 1.

5 Evaluation

We present the evaluation results of AudioIO in this section. The collected data samples are processed in an offline way using MATLAB and LibSVM. We take the indoor label as true in the training and testing phases.

5.1 Effect of Noise

Firstly, we evaluate the performance of the models on different noise levels. The chirp and MLS data sets are divided into three subsets respectively according to the background noise level: quiet (no noise generated), medium (the volume of generated white noise is 40 dB), and loud (the volume of generated white noise is 50 dB). We conduct 5-folds cross validation on each subset. It can be clearly observed from Fig. 6 that, the background noise does have a negative effect on the performance for both MLS and chirp signals. In the loud noise case, the performance decreases quite considerably compared with medium and quiet cases. But in the worst case AudioIO still achieves the accuracy of 92.17% (87.5% precision, 95.01% recall) with MLS signal, and the accuracy of 93.8% (93.75% precision, 98.02% recall) with chirp signal. Meanwhile, it can also be observed that there is no big difference between the performance of quiet and medium cases. It reveals that AudioIO is robust to background noise as long as it is not too noisy.

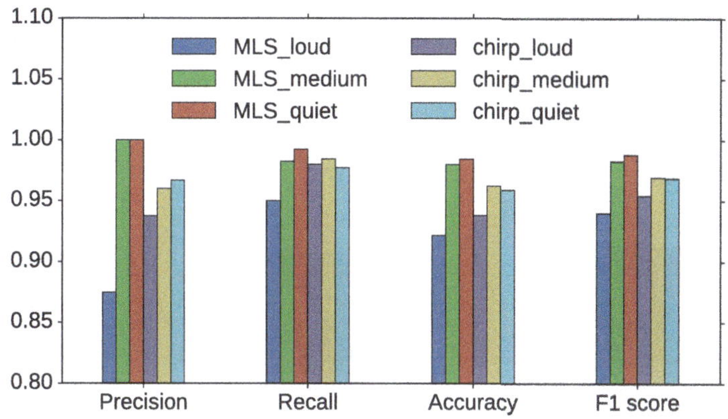

Fig. 6 Effect of noise level. The models are trained and tested on the data sets of all the devices based on 5-folds cross validation

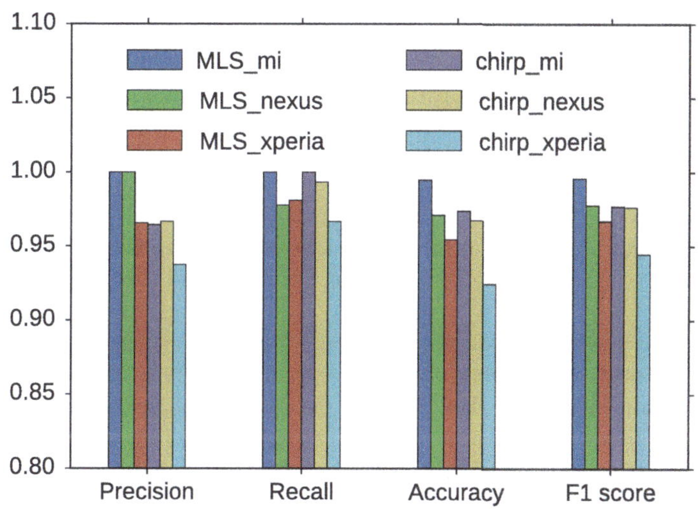

Fig. 7 Effect of devices. The models are trained and tested on the data sets of all noise levels based on 5-fold cross validation

5.2 Effect of Devices

Secondly we investigate into the effect of devices. Each pair of speaker and microphone has its own frequency property, which shapes the emitted signals and the measured samples. As a result, the audio responses of indoor/outdoor environments are also influenced. The chirp and MLS data sets are divided into three subsets respectively according to the devices: Mi, Nexus, and Xperia. Each category contains data samples of all noise levels. The 5-folds cross validation results in Fig. 7 show that the device type does have certain influence on the performance of

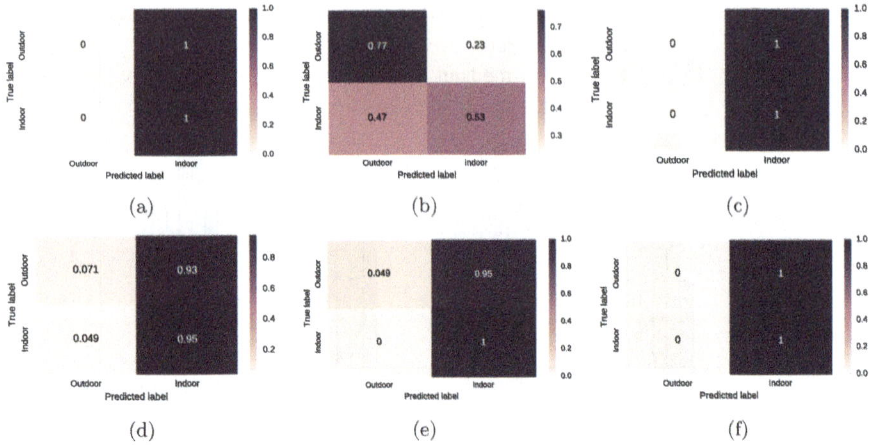

Fig. 8 Confusion Matrices for cross training on different devices. The models are trained on one device and then tested on the combined data set of the other two devices. For instance, *MLS: Trained on Mi* means the model is trained on the MLS signal data set on Mi, and tested on the data set of Nexus and Xperia. (**a**) Chirp: trained on Mi. (**b**) Chirp: trained on Nexus. (**c**) Chirp: trained on Xperia. (**d**) MLS: trained on Mi. (**e**) MLS: trained on Nexus. (**f**) MLS: trained on Xperia

AudioIO. Generally speaking AudioIO performs best on Mi and worst on Xperia, and MLS performs better than chirp. In the worst case, AudioIO still achieves the accuracy of 95.43% (96.55% precision, 98.1% recall) on Xperia with MLS signal, the accuracy of 92.46% (93.75% precision, 96.68% recall) on Xperia with chirp signal.

Further we conduct cross training to see if a model trained on one device can still work on other devices. For instance, we train a SVM classifier on the data set Mi with all noise levels and then test it on the data sets of Nexus and Xperia. The confusion matrices are shown in Fig. 8. It can be observed that the model trained on one device does not work on other devices. The reason is two-fold: firstly each smartphone has its own layout of speaker and microphone, which influences the received audio signal. Secondly the speaker and microphone on every device have their own frequency property, this property shapes the MFCC extracted from the IO reverberation differently. Without these prior knowledge, the learned model cannot transfer to a new device.

5.3 Overall Performance

Finally, we evaluate the performance of SVM classifiers trained on MLS and chirp data sets respectively. Both data sets consist of data samples of all noise levels and devices. As shown in Fig. 9, chirp signal outperforms MLS signal regarding precision, recall, accuracy, and F1 score. The SVM classifier trained on chirp data

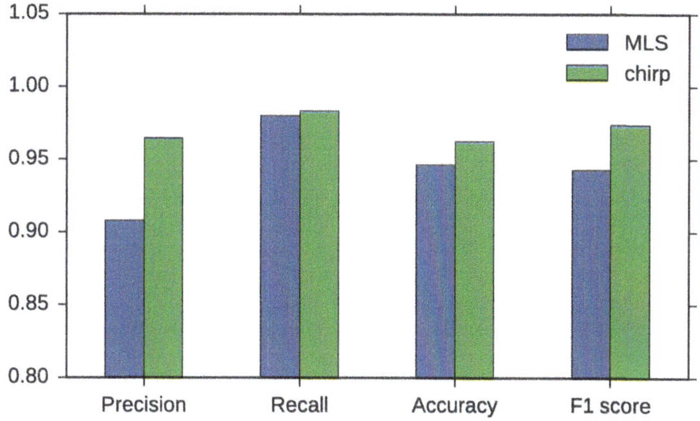

Fig. 9 Results of 5-folds cross validation for SVM models trained on both MLS and chirp signal data sets. The models are trained and tested on the data sets of all the devices and noise levels

set achieves the accuracy of 96.26% (96.47% precision, 98.34% recall). The chirp signal in our experiments lasts only 100 ms, which is much shorter than that of MLS signal (743 ms). It is more energy efficient to use chirp signal for active sound probing. Beside that the sound of chirp signal is less annoying than MLS signal. As a result, chirp signal fits better for the indoor/outdoor detection than MLS signal.

6 Conclusion and Discussion

We present an active sound probing based Indoor Outdoor detection method for smartphones. The embedded speaker and microphone are utilized to emit probing signal and record reverberations of the environments. Audio features are extracted from received signal and then fed to SVM classifier. We comprehensively evaluate Audio with different noise levels and device types. Results suggest that AudioIO achieves prompt and accurate detection results in various settings.

In indoor environments the reverberation profile consists of strong echoes from ceiling, ground, walls and weak echoes from neighboring objects. On the other hand in outdoor environments the strong echoes are less compared with indoor cases, which results in a different reverberation profile of the space. Though we achieved good results, AudioIO may fail if a room is full of textile fabrics such as carpet, tapestry which can absorb the probing signal. Additionally, the probing signal we use in the experiment is still hearable. Using the chirp signal in ultrasound frequency band is one interesting topic to be explored. Further we have not evaluated the energy efficiency of AudioIO, which we also reserve for future work.

References

1. Ali, M., ElBatt, T., Youssef, M.: SenseIO: realistic ubiquitous indoor outdoor detection system using smartphones. IEEE Sens. J. **18**(9), 3684–3693 (2018)
2. Amft, O., Van Laerhoven, K.: What will we wear after smartphones? IEEE Pervasive Comput. **16**(4), 80–85 (2017)
3. Beritelli, F., Grasso, R.: A pattern recognition system for environmental sound classification based on MFCCs and neural networks. In: 2nd International Conference on Signal Processing and Communication Systems, ICSPCS 2008, pp. 1–4. IEEE, Piscataway (2008)
4. Canovas, O., Lopez-de Teruel, P.E., Ruiz, A.: Detecting indoor/outdoor places using wifi signals and adaboost. IEEE Sens. J. **17**(5), 1443–1453 (2017)
5. Carroll, A., Heiser, G., et al.: An analysis of power consumption in a smartphone. In: USENIX Annual Technical Conference, Boston, vol. 14, pp. 21–21 (2010)
6. Chang, C.C., Lin, C.J.: LIBSVM: a library for support vector machines. ACM Trans. Intell. Syst. Technol. **2**(3), 27 (2011)
7. Chen, Y., Yonezawa, T., Nakazawa, J., Tokuda, H.: Evaluating the spatio-temporal coverage of automotive sensing for smart cities. In: 2017 Tenth International Conference on Mobile Computing and Ubiquitous Network (ICMU), pp. 1–5. IEEE, Piscataway (2017)
8. Chintalapudi, K., Padmanabha Iyer, A., Padmanabhan, V.N.: Indoor localization without the pain. In: Proceedings of the Sixteenth Annual International Conference on Mobile Computing and Networking, pp. 173–184. ACM, New york (2010)
9. Cho, H., Song, J., Park, H., Hwang, C.: Deterministic indoor detection from dispersions of GPS satellites on the celestial sphere. In: The 11th International Symposium on Location Based Services (2014)
10. Fan, M., Adams, A.T., Truong, K.N.: Public restroom detection on mobile phone via active probing. In: Proceedings of the 2014 ACM International Symposium on Wearable Computers, pp. 27–34. ACM, New York (2014)
11. Franke, T., Lukowicz, P., Blanke, U.: Smart crowds in smart cities: real life, city scale deployments of a smartphone based participatory crowd management platform. J. Internet Serv. Appl. **6**(1), 27 (2015)
12. Ishida, Y., Thepvilojanapong, N., Tobe, Y.: Winfo+: identification of environment condition using walking signals. In: Tenth International Conference on Mobile Data Management: Systems, Services and Middleware, MDM'09, pp. 508–512. IEEE, Piscataway (2009)
13. Jia, M., Yang, Y., Kuang, L., Xu, W., Chu, T., Song, H.: An indoor and outdoor seamless positioning system based on android platform. In: 2016 IEEE Trustcom/BigDataSE/ISPA, pp. 1114–1120. IEEE, Piscataway (2016)
14. Khaled, A.E., Helal, A., Lindquist, W., Lee, C.: IoT-DDL-device description language for the "T" in IoT. IEEE Access **6**, 24048–24063 (2018)
15. Lipowezky, U., Vol, I.: Indoor-outdoor detector for mobile phone cameras using gentle boosting. In: 2010 IEEE Computer Society Conference on Computer Vision and Pattern Recognition Workshops (CVPRW), pp. 31–38. IEEE, Piscataway (2010)
16. Maeda, H., Sekimoto, Y., Seto, T.: An easy infrastructure management method using on-board smartphone images and citizen reports by deep neural network. In: Proceedings of the Second International Conference on IoT in Urban Space, pp. 111–113. ACM, New York (2016)
17. Maeda, H., Sekimoto, Y., Seto, T.: Lightweight road manager: smartphone-based automatic determination of road damage status by deep neural network. In: Proceedings of the 5th ACM SIGSPATIAL International Workshop on Mobile Geographic Information Systems, pp. 37–45. ACM, New York (2016)
18. Nakamura, Y., Ono, M., Sekiya, M., Honda, K., Takahashi, O.: Indoor/outdoor determination method using various sensors for the power saving of terminals in geo-fencing. In: Proceedings of the 2015 International Workshop on Informatics (2015)

19. Okamoto, M., Chen, C.: Improving GPS-based indoor-outdoor detection with moving direction information from smartphone. In: Adjunct Proceedings of the 2015 ACM International Joint Conference on Pervasive and Ubiquitous Computing and Proceedings of the 2015 ACM International Symposium on Wearable Computers, pp. 257–260. ACM, New York (2015)
20. Oppenheim, A.V., Willsky, A.S., Nawab, S.H.: Signals and Systems, vol. 2. Prentice-Hall, Englewood Cliffs (1983). **6**(7), 10
21. Perttunen, M., Mazhelis, O., Cong, F., Kauppila, M., Leppänen, T., Kantola, J., Collin, J., Pirttikangas, S., Haverinen, J., Ristaniemi, T., et al.: Distributed road surface condition monitoring using mobile phones. In: International Conference on Ubiquitous Intelligence and Computing, pp. 64–78. Springer, Berlin (2011)
22. Radu, V., Katsikouli, P., Sarkar, R., Marina, M.K.: A semi-supervised learning approach for robust indoor-outdoor detection with smartphones. In: Proceedings of the 12th ACM Conference on Embedded Network Sensor Systems, pp. 280–294. ACM, New York (2014)
23. Rossi, M., Feese, S., Amft, O., Braune, N., Martis, S., Tröster, G.: AmbientSense: a real-time ambient sound recognition system for smartphones. In: 2013 IEEE International Conference on Pervasive Computing and Communications Workshops (PERCOM Workshops), pp. 230–235. IEEE (2013)
24. Rossi, M., Seiter, J., Amft, O., Buchmeier, S., Tröster, G.: Roomsense: an indoor positioning system for smartphones using active sound probing. In: Proceedings of the 4th Augmented Human International Conference, pp. 89–95. ACM, New York (2013)
25. Shtar, G., Shapira, B., Rokach, L.: Clustering wi-fi fingerprints for indoor-outdoor detection. Wirel. Netw. **25**(3), 1341–1359 (2018)
26. Stan, G.B., Embrechts, J.J., Archambeau, D.: Comparison of different impulse response measurement techniques. J. Audio Eng. Soc. **50**(4), 249–262 (2002)
27. Sung, R., Jung, S.H., Han, D.: Sound based indoor and outdoor environment detection for seamless positioning handover. ICT Express **1**(3), 106–109 (2015)
28. Tahir, W., Majeed, A., Rehman, T.: Indoor/outdoor image classification using gist image features and neural network classifiers. In: 2015 12th International Conference on High-Capacity Optical Networks and Enabling/Emerging Technologies (HONET), pp. 1–5. IEEE, Piscataway (2015)
29. Uehara, Y., Mori, M., Ishii, N., Tobe, Y., Shiraishi, Y.: Step-wise context extraction in aok mule system. In: Proceedings of the 4th International Conference on Embedded Networked Sensor Systems, pp. 379–380. ACM, New York (2006)
30. Wang, Y., Lin, J., Annavaram, M., Jacobson, Q.A., Hong, J., Krishnamachari, B., Sadeh, N.: A framework of energy efficient mobile sensing for automatic user state recognition. In: Proceedings of the 7th International Conference on Mobile Systems, Applications, and Services, pp. 179–192. ACM, New York (2009)
31. Wang, H., Sen, S., Elgohary, A., Farid, M., Youssef, M., Choudhury, R.R.: No need to war-drive: unsupervised indoor localization. In: Proceedings of the 10th International Conference on Mobile Systems, Applications, and Services, pp. 197–210. ACM, New York (2012)
32. Wang, W., Chang, Q., Li, Q., Shi, Z., Chen, W.: Indoor-outdoor detection using a smart phone sensor. Sensors **16**(10), 1563 (2016)
33. Zhou, P., Zheng, Y., Li, Z., Li, M., Shen, G.: IODetector: a generic service for indoor outdoor detection. In: Proceedings of the 10th ACM Conference on Embedded Network Sensor Systems, pp. 113–126. ACM, New York (2012)
34. Zou, H., Jiang, H., Luo, Y., Zhu, J., Lu, X., Xie, L.: BlueDetect: an iBeacon-enabled scheme for accurate and energy-efficient indoor-outdoor detection and seamless location-based service. Sensors **16**(2), 268 (2016)

Day-Ahead Load Forecasting Based on Conditional Linear Predictions with Smoothed Daily Profile

Sunme Park, Kanggu Park, and Euiseok Hwang

1 Introduction

Global energy demand has recently been on the rise owing to strong global economic growth and a faster increase in electricity consumption than in other sectors as based on the electrification of energy consumption [1]. In particular, buildings account for more than 55% of global electricity use, which is leading the growth in energy demand [2]. To manage this significant increase in power demand while preserving environmental factors, smart grid technologies are actively being investigated and deployed in various ways, including the use of advanced energy management systems (EMSs) for providing sufficient amounts of electricity with efficient and stable power system operations [3]. In a smart grid, prior knowledge regarding electric power consumption is important in coordinating the grid in terms of efficiency and sustainability by reducing uncertainty and dealing with potential issues that might arise. In particular, load forecasting is a fundamental technique for planning an effective smart grid operation and preventing potential blackouts by stabilizing the power system. In addition, a forecasted load is also important on the demand side when participating in a demand response program, which operates as a distributed energy resource (DER) relieving any problems with the grid [4, 5]. This study focuses on the short-term load forecasting (STLF), particularly day-ahead prediction, which is essential for reducing energy costs or executing a demand response for the following day by allowing consumers to customize their energy use patterns based on prior information.

S. Park · K. Park · E. Hwang (✉)
Gwangju Institute of Science and Technology, Gwangju, South Korea
e-mail: pishp00200@gist.ac.kr; kgpark09@gist.ac.kr; euiseokh@gist.ac.kr

R. José et al. (eds.), *3rd EAI International Conference on IoT in Urban Space*, EAI/Springer Innovations in Communication and Computing,
https://doi.org/10.1007/978-3-030-28925-6_9

A conventional time-series model of the load prediction may encounter a refinement of explanatory data containing random fluctuations. Thus, in the present study, a pre-processing of the input sequences, namely, the records of the previous day, is needed for effectively eliminating sequential randomness. In this paper, data binning is proposed as a pre-processing technique for electricity data, where the number of coefficients can be significantly reduced concurrently through a set of linear prediction filters. The performance of the proposed binning for day-ahead load prediction is evaluated using campus load data from a 3.5-year record of 15-min samples. The results show that smoothing the input load profile can improve the prediction performance as compared to the use of a full record of raw data. The number of bins can be determined based on a favorable trade-off between extracting the input and a loss of information.

The remainder of this paper is organized as follows. Section 2 introduces related studies conducted on data-driven models, particularly the statistical approach of load prediction. In Sect. 3, the conditional linear regression model for day-ahead load prediction is described. In Sect. 4, a correlation-based analysis is presented for verifying the data binning effect. The load prediction results of the proposed scheme are then evaluated in Sect. 5, and finally Sect. 6 provides some concluding remarks regarding this research.

2 Related Works

The building of load forecast models can be categorized into three approaches. First, white model forecasting, which is also known as mathematical modeling, is established based on the physical components of a building. Next, a gray box model combines a data-based model with partial information of a single system. Finally, black box modeling is a data-driven model developed from a historical dataset [7]. A data-driven approach is effective when the physical data are shortened. For load forecasting, the most frequently used data-driven prediction models are time-series statistical regression, a decision tree (DT), artificial neural networks (ANNs), a support vector machine (SVM), and a genetic algorithm [6].

A statistical method is used to determine the energy consumption with a correlation between the various types of factors and the historical data. The similar-day approach is the earliest STLF method, with intuitive insight and a simple implementation [7]. In addition, a multiple regression model can reflect various properties of external information. Because the energy consumption depends on the weather and daily routine of the occupants [8], for load forecasting, weather data such as the temperature, wind speed, and humidity, and the day of the week or time index, are commonly considered factors. The input features strongly related to electricity usage can be selected for a prediction [9] by normalizing their impact when the ambient temperature, weekday index, and building types are selected for a load prediction using a multiple regression model.

More specifically, linear regression combined with a time-series analysis has been widely proposed in many different studies. The most popular methods are an autoregressive (AR) approach and the use of the moving average (MA), whose integration has been applied to the differencing process named as ARIMA. This model is based on the idea of stationary data, which is expressed through a linear regression of the time-series data [10]. Alternatively, multiple seasonal factors are considered, namely, intraday, intraweek, and intrayear, and are applied to the seasonal ARIMA and exponential smoothing, which shows that the combination of these methods greatly improves the forecasting accuracy by lowering the measurement errors [11]. In addition, several studies have applied the pre-processing of the input data prior to the forecasting load. Linear regression is conducted by eliminating trends and seasonal effects, allowing a typical daily profile to be utilized as an input, thereby keeping the model as simple as possible, resulting in a better performance over conventional STLF methods [12]. Exponential smoothing is another method widely used in load forecasting, and the modified Holt-Winters exponential smoothing is applied to reflect the double seasonality in half hourly data [13].

3 Day-Ahead Load Prediction Based on Smoothed Profile of Previous Day

3.1 Conditional Linear Prediction

Electricity demand is affected by various factors—temperature, trend, economic conditions, and the social component associated with a calendar. Especially for day-ahead prediction, it is essential of calendar day to be distinguished into day type, e.g., $h_d \in \{W, N\}$ denoting workday and non-workday, respectively, whereas the effect of the other factors is smooth and can be assumed to be similar for neighboring days. With consideration of this property, day type conditioned linear prediction is developed in this study for day-ahead load forecasting using previous day observations. The conditional linear prediction model can be revised in matrix form as following equation:

$$y_{t,d} = \mathbf{a}_{t,h_d|h_{d-1}}^T \cdot \mathbf{y}_{d-1}^{(n)} + \epsilon_{t,d} \tag{1}$$

where $y_{t,d}$ is the load of time t of day d, and $\mathbf{y}_{d-1}^{(n)}$ is the load of the previous day with length n, e.g., 96 for a 15-min sampled sequence. The linear relation is captured using $\mathbf{a}_{t,h_d|h_{d-1}}^T$ for the designated time t for a day of type h_d followed by a h_{d-1} type day, and $\epsilon_{t,d}$ is the residue for any non-linear contributions. The model can be represented as follows:

$$\mathbf{y}_d^{(n)} = \mathbf{A}_{h_d|h_{d-1}}^T \cdot \mathbf{y}_{d-1}^{(n)} + \epsilon_d \tag{2}$$

where $A_{h_d|h_{d-1}}$ is an $n \times n$ matrix whose t-th column vector is $\mathbf{a}_{t,h_d|h_{d-1}}$. Multiple seasonal patterns such as weekly or annual cycles are filtered by the given day type conditions, which leads to a lighter equation, whereas conventional studies gradually add seasonal terms.

$\mathbf{y}$	daily load	d	day index
$\mathbf{y}$	daily load with binning	h	daytype
$\mathbf{A}$	linear prediction coefficient matrix	t_1	time point of previous day
$\mathbf{R}$	correlation matrix	t_2	time point of target day
ρ	Pearson coefficient	trn	training
$\hat{}$	estimated value	tst	test

The linear prediction coefficient (LPC) can be obtained through the linear minimum mean square error (LMMSE) using the following equation:

$$\hat{y}_{t,d} = \hat{\mathbf{a}}^T_{t,h_d|h_{d-1}} \cdot \mathbf{y}^{(n)}_{d-1} \tag{3}$$

$$\hat{\mathbf{a}}_{t,h_d|h_{d-1}} = \left(\hat{\mathbf{R}}_{\mathbf{y}^{(n)}_{d-1}|h_d,h_{d-1}}\right)^{-1} \cdot \hat{\mathbf{R}}_{\mathbf{y}^{(n)}_{d-1}y_{t,d}|h_d,h_{d-1}} \tag{4}$$

where

$$\hat{\mathbf{R}}_{\mathbf{y}^{(n)}_{d-1}|h_d,h_{d-1}} = \hat{E}[\mathbf{y}^{(n)}_{d-1}\mathbf{y}^{(n)}_{d-1}{}^T|h_d, h_{d-1}] \tag{5}$$

is an $n \times n$ empirical correlation matrix of the previous day profile and

$$\hat{\mathbf{R}}_{\mathbf{y}^{(n)}_{d-1}y_{t,d}|h_d,h_{d-1}} = \hat{E}[\mathbf{y}^{(n)}_{d-1}y_{t,d}|h_d, h_{d-1}] \tag{6}$$

is an $n \times 1$ cross-correlation vector between the previous profile and the desired output at time t on the target day d. In addition, $\hat{E}[\cdot]$ denotes an empirical correlation estimate from the historical days, the variance of which depends on the number of corresponding days of the pair (h_d, h_{d-1}) in the records.

3.2 *Data Binning*

Data binning is a simple data pre-processing method used to reduce the effects of minor observational errors. Quarter-hourly sampled load data has minor fluctuations and noise that degrade the linear regression modeling used for a prediction because their variety weakens the correlation between the input and output data. To mitigate the uncertainty, data binning is proposed in which the input records of the previous day are grouped into m bins of varying size b_k and averaged, as illustrated in Fig. 1. Data binning can also reduce the computational complexity compared to a model using all points by decreasing the LPC filter length of the linear regression model. Using binning, the linear prediction in Eqs. (3)–(6) can be modified by replacing $\mathbf{y}^{(n)}_{d-1}$ with $\bar{\mathbf{y}}^{(m)}_{d-1}$.

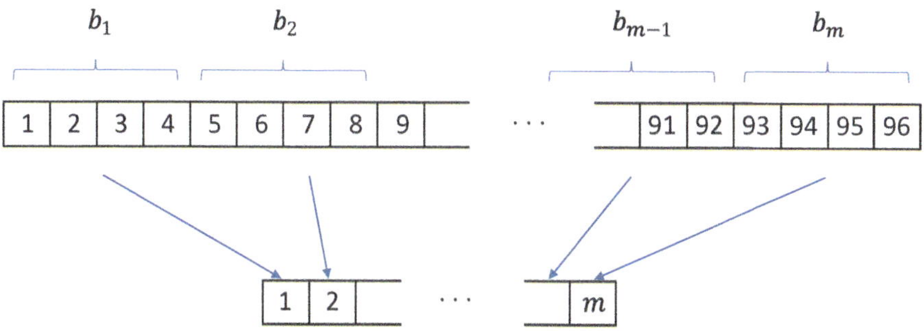

Fig. 1 Illustration of data binning for daily profile of $n = 96$ points by grouping into size b_k for each bin for $k = 1, 2, \ldots, m$

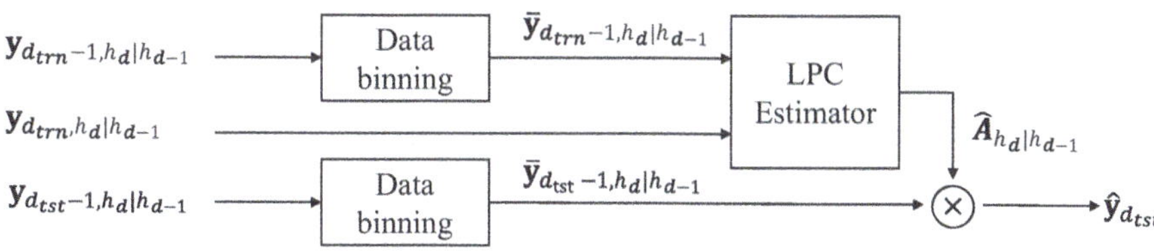

Fig. 2 Scheme of the proposed system

$$\bar{y}_{t',d-1} = \frac{1}{b_{t'}} \Sigma_{j=1}^{b_{t'}} y_{\Sigma_{k=1}^{t'-1}(b_k)+j,d-1} \tag{7}$$

where $t' = 1, 2, \ldots, m$, the number of bins m is identical to the LPC filter length, and b_k represents the bin size. Note that the correlation matrix and vector in Eqs. (5) and (6) become $m \times m$ and $m \times 1$, respectively, through the use of binning, whereas the number of LPC filters in Eqs. (3) and (4) is the same as n. For binning of a regular size, all bins are of equal size, that is

$$b_1 = b_2 = \cdots = b_m, \ \Sigma_{k=1}^{m} b_k = 96 \tag{8}$$

For a performance evaluation, the input dataset of twelve different sized bins is tested to predict 96 points under different numbers of bins ($m = 1, 2, 3, 4, 6, 8, 12, 16, 24, 32, 48, 96$).

3.3 System Scheme

During the training phase, the LPC is estimated from historical data pairs $(\bar{\mathbf{y}}_{d_{trn}-1}, \mathbf{y}_{d_{trn}})$. Such pairs are collected until the day before the test day by dividing them into individual types of day. To estimate $\hat{\mathbf{y}}_{d_{tst}}$, the LPC obtained is multiplied by the binning profile of the previous day $\bar{\mathbf{y}}_{d_{tst}-1}$ (Fig. 2).

4 Campus Load Data Analysis

4.1 Monitored Data

For the evaluation, the electricity demand for a 3.5-year period was monitored at Gwangju Institute of Science and Technology (GIST) in South Korea (January 2015 to June 2018). The demand was measured every 15 min for each daily profile. As indicated in Fig. 3a, the upper trend of the electricity demand over the year was observed as new buildings and equipment were constructed. In addition, it is important to note the annual seasonality because of the heating, ventilating, and air conditioning (HVAC) usage associated with the weather conditions. Because electricity demand is mainly affected by the social aspects of human activities, the daily patterns, as shown in Fig. 3b, were captured based on the workday profile, which is higher than the non-workday profile.

4.2 Correlation Analysis

With a linear regression model, the correlation has a direct influence on the building forecasting model. To analyze the cross-correlation between the load data of the previous day $\mathbf{y}_{t_1,h_{d-1}}$ and that of the target day $\mathbf{y}_{t_2,h_d}$, a Pearson's coefficient was

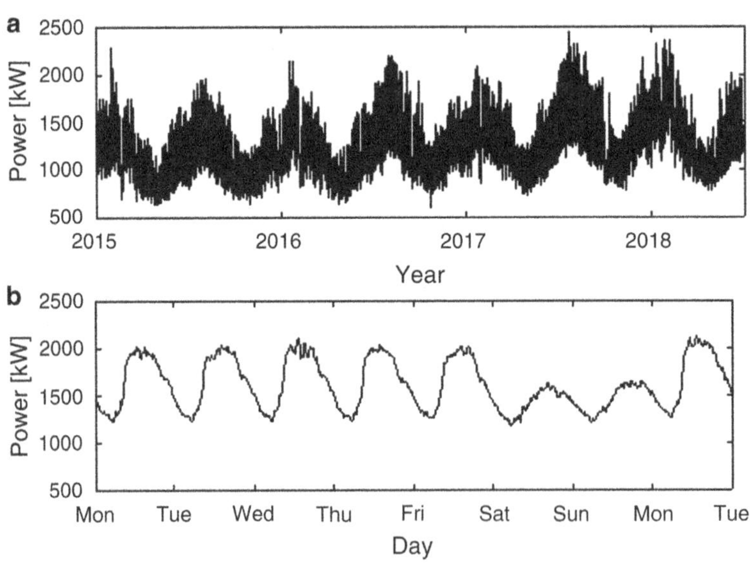

Fig. 3 GIST campus electricity data monitored (**a**) from January 2015 to June 2018 and (**b**) for 1-week

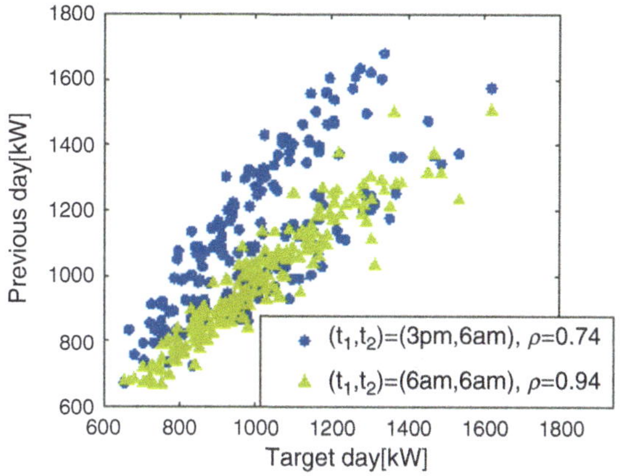

Fig. 4 Load demand scatter plotted. When forecasting $t_2 = 6$ am on (h_{d-1}, h_d) =(N, W), the correlation of (t_1, t_2) = (6 am,6 am) is higher than that of (t_1, t_2) = (3 pm,6 am), which means that $t_1 = 6$ am data are important for forecasting $t_2 = 6$ am on a workday

derived, as shown in Eq. (9), when considering four types of day, namely, (h_{d-1}, h_d) =(W, W), (N, W), (W, N), (N, N).

$$\rho_{\mathbf{y}_{t_1,h_d},\mathbf{y}_{t_2,h_{d-1}}} = \frac{cov(\mathbf{y}_{t_1,h_d}, \mathbf{y}_{t_2,h_{d-1}})}{\sigma_{\mathbf{y}_{t_1,h_d}} \cdot \sigma_{\mathbf{y}_{t_2,h_{d-1}}}} \tag{9}$$

Here, t_1 is the index of the previous day, which is determined based on the binning size $(t_1 = 1, 2, \ldots, n)$, and t_2 is the time index of the target day $(t_2 = 1, 2, \ldots, 96)$.

Figure 4 shows an example of obtaining a correlation coefficient. All datasets correspond to the type of day (N,W), where case 1, namely, (t_1, t_2) = (6 am,6 am), and case 2, namely, (t_1, t_2) = (3 pm, 6 am), are scatter plotted. Case 1 shows a high correlation of $\rho = 0.94$ because the dataset is concentrated along a line, whereas case 2 has relative sparsity, leading to a low correlation of $\rho = 0.74$. In other words, when building a forecasting model of (h_{d-1}, h_d) = (N,W), the data from the previous day at 6 am are more deterministic for forecasting at 6 am than are the data of the previous day at 3 pm.

Figure 5a shows the correlation between the binning index of t_1 and the quarter-hourly sampled t_2 divided into all types of day. The degree of correlation is indicated based on color, and is formed into an $m \times 96$ matrix called a correlation map. The higher the correlation is, the brighter the color used in the correlation map. The map draws a high correlation in an "X" shape because the load of the target day is in general highly correlated with the same time on the previous day regardless of the type of day. In addition, the time slots for going to and leaving work have a low correlation with the other time slots, which is considered to be based on the fact that, during these time periods, the occupants use a constant amount of energy despite the seasonality, thereby indicating that the dataset has a small covariance.

As shown in Fig. 5b, the bright areas of the correlation map are increased, mitigating the low covariance effect. The number of correlation coefficients of larger than 0.95 for each correlation map is determined and indicated as a ratio, as shown

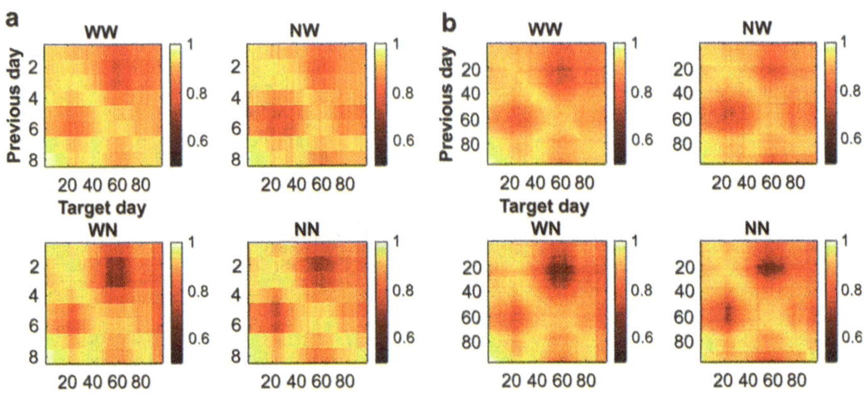

Fig. 5 Correlation map between 15-min sampled loads of the target day and m-bins of the previous day. Here, (**a**) $m = 96$, namely, all points, forming a 96×96 correlation map, and (**b**) $m = 8$, forming an 8×96 correlation map for each type of day

Table 1 Ratio of correlation coefficient of larger than 0.95 for each dataset

(h_{d-1}, h_d)	(W,W)	(N,W)	(W,N)	(N,N)
Full points	3.28	1.56	3.34	2.25
8-binning	5.86	2.86	5.08	4.04

Eight-bin data show a general highly correlated proportion compared to the full-point data for all types of day (h_{d-1}, h_d)

in Table 1. The eight-bin data of the previous day, which are highly correlated with the target dataset, demonstrate a large rate compared to the full-point data used for all types of day.

5 Performance Evaluations

To evaluate the forecasting performance, the 3.5-year dataset of the campus load was divided into a 2.5-year training set and a 1-year test set (from July 2017 to June 2018). The mean absolute error (MAE) and mean absolute percentage error (MAPE) are applied for an accuracy evaluation, and their standard deviation (STD) is used to evaluate the stability of the model.

$$\mathrm{MAE[kW]} = \frac{1}{n}\Sigma_{t=1}^{n}|\hat{y}_{t,d} - y_{t,d}| \tag{10}$$

$$\mathrm{MAPE}\ [\%] = \frac{1}{n}\Sigma_{t=1}^{n}|\hat{y}_{t,d} - y_{t,d}|/y_{t,d} * 100 \tag{11}$$

where $y_{t,d}$ and $y_{t,d}$ denote the predicted load and observed value, respectively, at time index t on target day d, and n indicates the total number of forecasting points, i.e., $n = 96$.

5.1 Data Binning

As presented in Fig. 6, through the binning process, a smoothing effect can be yielded compared to the full 96-point profile. This can decrease the load variation and avoid an undesired correlation effect. However, a trade-off exists between the smoothing effect and information loss, accompanied with the binning process.

To construct a linear regression model from historical data, it is important to determine how many days are used for the training period. Because linear regression requires a large amount of historical data, we can resolve the risk of a lack of data by testing various sized training sets. Therefore, we tested different training periods using a sliding window focusing on the recent data for each test day. The performance when testing under various window sizes (1–30 months), and applying an accumulated window (AW) to all recorded data until the day prior to the test day, is shown in Fig. 7.

It was revealed that the training period lengthens as the number of bins increases because a large number of data points require a large dataset when determining the LPC. A training period with a window size of 18–24 months achieves a slightly lower performance when compared to the neighboring window size. During this period, data from the opposite season aggravate the forecasting performance when reflecting the different annual seasons. Figure 8 indicates that the performance improves with the amount of binning data in accordance with the smoothing effect. However, a small number of bins generate a large amount of information loss, resulting in low accuracy.

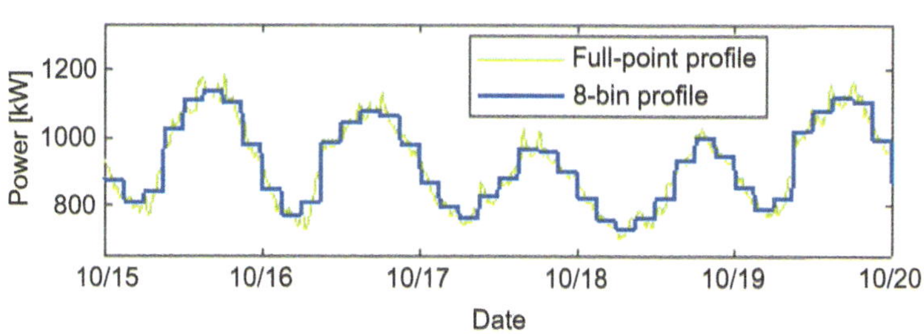

Fig. 6 Full 96-point profiles and eight-bin processed profile. The binning profile decreases the fluctuation by grouping and averaging in each bin.

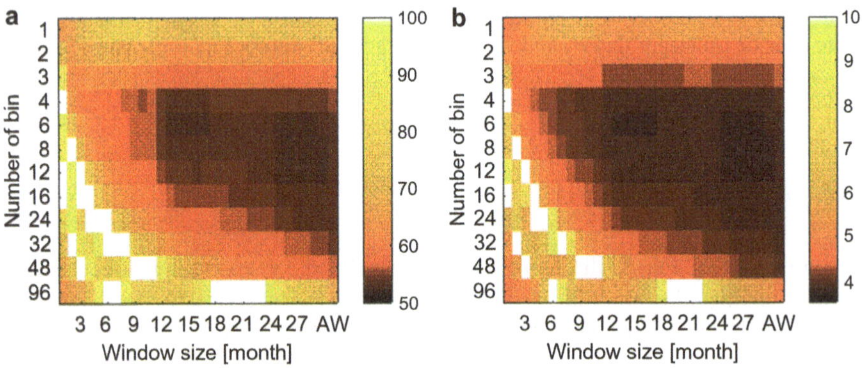

Fig. 7 Forecasting error of MAE (**a**) and MAPE (**b**) when varying the number of bins and the window size

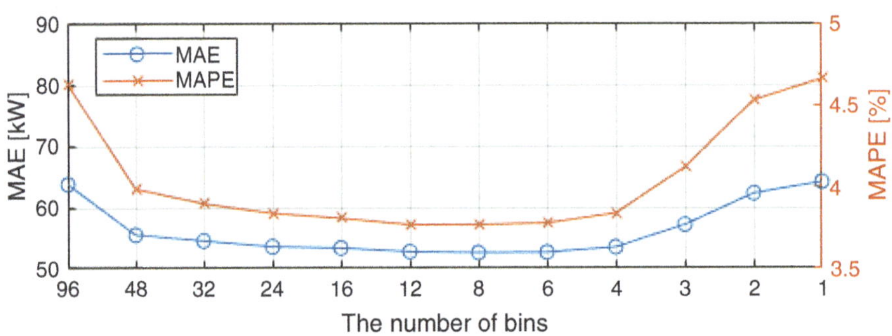

Fig. 8 Forecasting error for different numbers of bins. Binning-based prediction leads to low errors in accordance with the smoothing effect. However, a small set of bins causes information loss, resulting in low accuracy

5.2 *Day-Ahead Prediction Performance*

For day-ahead forecasting of GIST campus electricity demand, we chose eight-bin data with a 26-month fixed sliding training set for comparison with full 96-point based prediction, as per the results shown in Fig. 8. Binning-based forecasting achieves a better real value than full 96-point based forecasting on both workdays and non-workdays. As shown in Table 2, the binning-based forecasting performance is improved as compared to unprocessed forecasting. Whereas the MAE and MAPE for full 96-point based forecasting are 4.63% and 63.85 kW, respectively, and the binning-based forecasting results are 3.77% and 52.64 kW, respectively. The STDs of the daily MAPE and MAE are additionally considered to verify the model stability. The STDs of both MAE and MAPE are reduced by 29.18 kW and 1.85% with the binning-based model, which means the proposed binning method achieves less variability in the daily load forecasting (Fig. 9).

Table 2 Evaluation of accuracy and stability of the forecasting model with and without binning

	Full-point pred.	Eight-bin pred.
MAE [kW]	63.85	52.64
(STD)	(35.70)	(29.18)
MAPE [%]	4.63	3.77
(STD)	(2.45)	(1.85)

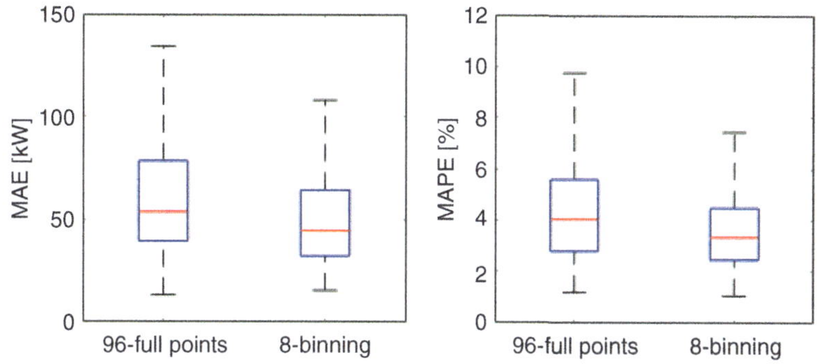

Fig. 9 Box plot of forecasting error comparing full-point and eight-bin based predictions. The binning-based model improves the prediction accuracy on average with less variability

6 Conclusion

In this study, a binning-based day-ahead short-term load prediction was proposed in which the records from previous days are smoothed for linear prediction. Day type conditioning is necessary for a day-ahead prediction, where random fluctuations can be severed using the full-day sequence, such as 96 data points through quarterly-hour sampling. A dataset of campus electricity use for a 3.5-year period was evaluated for a correlation analysis and prediction testing. Various sizes of regular binning were tested, and a dimensional reduction through binning was demonstrated to provide favorable correlation measures on the same day. In addition, day-ahead forecasting based on the conditional linear prediction performance was also shown to generally improve through binning as based on the MAE and MAPE measures. When using eight bins, both the MAE and MAPE were improved by approximately 20% on average for 1 year tests, as compared to using a full set of 96 points. In addition, the deviation in the prediction errors was measured to investigate the stability of daily forecasting, where binning-based prediction demonstrated an approximately 20% smaller deviation than full-point prediction. This indicates that the proposed binning-based prediction shows a potential improvement of load forecasting performance in terms of both accuracy and stability.

Acknowledgements This work was supported by GIST Research Institute (GRI) grant funded by the GIST in 2018 and by the Korea Institute of Energy Technology Evaluation and Planning (KETEP) and the Ministry of Trade, Industry and Energy (MOTIE) of the Republic of Korea (No. 20171210200810).

References

1. Global Energy Statistical Yearbook. Enerdata (2018). https://yearbook.enerdata.net. Accessed on September 2018
2. Energy technology perspectives 2017: Catalysing energy technology transformations, IEA (2017)
3. Zhang, Y., Mao, M., Ding, M., Chang, L.: Study of energy management system for distributed generation systems. In: Third International Conference on Electric Utility Deregulation and Restructuring and Power Technologies, DRPT 2008, pp. 2465–2469. IEEE, Piscataway (2008)
4. Siano, P.: Demand response and smart grids: a survey. Renew. Sustain. Energy Rev. **30**, 461–478 (2014)
5. O'Connell, S., Riverso, S.: Flexibility analysis for smart grid demand response (2017). Preprint arXiv:1704.01308
6. Wei, Y., Zhang, X., Shi, Y., Xia, L., Pan, S., Wu, J., Han, M., Zhao, X.: A review of data-driven approaches for prediction and classification of building energy consumption. Renew. Sustain. Energy Rev. **82**, 1027–1047 (2018)
7. Hong, T., Gui, M., Baran, M.E., Willis, H.L.: Modeling and forecasting hourly electric load by multiple linear regression with interactions. In: 2010 IEEE Power and Energy Society General Meeting, pp. 1–8. IEEE, Piscataway (2010)
8. Yarbrough, I., Sun, Q., Reeves, D.C., Hackman, K., Bennett, R., Henshel, D.S.: Visualizing building energy demand for building peak energy analysis. Energy Build. **91**, 10–15 (2015)
9. Amber, K.P., Aslam, M.W., Mahmood, A., Kousar, A., Younis, M.Y., Akbar, B., Chaudhary, G.Q., Hussain, S.K.: Energy consumption forecasting for university sector buildings. Energies **10**(10), 1579 (2017)
10. Deb, C., Zhang, F., Yang, J., Lee, S.E., Shah, K.W.: A review on time series forecasting techniques for building energy consumption. Renew. Sustain. Energy Rev. **74**, 902–924 (2017)
11. Taylor, J.W.: Triple seasonal methods for short-term electricity demand forecasting. Eur. J. Oper. Res. **204**(1), 139–152 (2010)
12. Dudek, G.: Pattern-based local linear regression models for short-term load forecasting. Electr. Power Syst. Res. **130**, 139–147 (2016)
13. Jalil, N.A.A., Ahmad, M.H., Mohamed, N.: Electricity load demand forecasting using exponential smoothing methods. World Appl. Sci. J. **22**(11), 1540–1543 (2013)

GANonymizer: Image Anonymization Method Integrating Object Detection and Generative Adversarial Network

Tomoki Tanimura, Makoto Kawano, Takuro Yonezawa, and Jin Nakazawa

1 Introduction

Today's cities are captured by thousands of cameras. By collecting and analyzing image data from such ubiquitous urban cameras, we could understand and even predict various contexts of cities. The cameras in cities are basically divided into two types—the fixed camera such as a security camera, and the mobile camera such as a drive recorder. Since mobile cameras on vehicles or people's smartphones have great possibility to capture entire city, several research has been proposed to leverage image data from the cameras for investigating and understanding urban infrastructure such as road condition [7, 12] and disaster [10]. If we collect, share and use the image data taken by thousands of mobile cameras, a large image dataset of the city, which must be very useful for understanding the city, can be generated. However, the image data of urban cameras, especially mobile cameras, always faces the risk of privacy invasion since the data includes objects related to privacy such as pedestrians and cars. Thus, to avoid the privacy data leakage, image data which may have the risk of privacy invasion cannot be shared widely. For example, though Fujisawa city in Japan has hundreds of dashcam on public vehicles, taken images cannot be shared among different organizations or even the images are discarded for reducing the risk of privacy data leakage. In summary, IoT-ready urban cameras cannot share their valuable image data for the privacy reason. Anonymizing image data from urban cameras must be an approach to accelerate sharing the image data.

T. Tanimura (✉) · M. Kawano · T. Yonezawa · J. Nakazawa
Keio University, Fujisawa, Kanagawa, Japan
e-mail: tanimu@ht.sfc.keio.ac.jp; makora@ht.sfc.keio.ac.jp; takuro@ht.sfc.keio.ac.jp; jin@ht.sfc.keio.ac.jp

R. José et al. (eds.), *3rd EAI International Conference on IoT in Urban Space*, EAI/Springer Innovations in Communication and Computing,
https://doi.org/10.1007/978-3-030-28925-6_10

In this paper, we propose GANonymizer that automatically detects and removes objects related to privacy such as persons and cars from an image. GANonymizer composes two machine learning networks: a privacy detection network and an object removal network. The privacy detection network detects objects related to privacy such as persons and cars from an image. The object removal network that removes the detected objects naturally as though they are not exist originally. To combine these two networks, we developed an edge shift padding layer (ESP) and a global feature padding layer (GFP) between the two networks in order to adapt GANonymizer to the characteristics of the images of the city and make GANonymizer to remove objects more naturally. ESP helps GANonymizer to remove objects placed at the edge of an image and GFP helps GANonymizer to remove large objects in an image. In our experiment, we adapt GANonymizer to anonymize videos (5246 images) taken from a car running in Fujisawa city, Japan. Implemented GANonymizer detects and removes persons, cars, buses, bikes, and bicycles as the targets of objects related to privacy based on the features of the video. We confirmed that our method contributes to remove privacy-concern objects.

2 Privacy Protection for Urban Video Images

2.1 *Understanding City Context by Analyzing Urban Video Images*

By analyzing of a large dataset of urban video images, we can understand latest context of city. For example, Kawano et al. have proposed the method for automatically detecting blur lines and signs on the road [7]. They also argued that leveraging video image of dashcam on public city garbage trucks can cover entire area of city. Moreover, Maeda et al. have proposed the method for detecting a damage on a road [12]. They used the light model, MobileNet SSD as an object detection network and trained the model with the dataset annotated with damages on a road. Furthermore, they proposed the smartphone application that automatically detects damages on a road using its camera in real time and also opened the dataset used in the model training to the public. These methods of analyzing urban images enable us to automatically monitor urban infrastructure. Not only urban infrastructure context, but also seasonal changes can be analyzed by urban video images. For example, Morishita et al. proposed SakuraSensor which extracts and shares information of flowering cherries along roads by using smartphone camera on car [13]. These existing works presented usefulness of urban video images—we can extract various city context by sharing and analyzing urban video from many vehicles.

2.2 Privacy Problem When Sharing Urban Video Images

Though we can analyze video images on edge side and share only result of the analysis, some application scenarios such as efficient road damage management require to share video images themselves. For example, we interviewed to city officers of road management section in Fujisawa city, Japan about how automatic road damage detection technique [7, 12] could be useful for actual road management. They said that automatic road damage indexing of whole city is very useful, but it is still necessary to check actual road images by human eyes for deciding which road should be repaired. Thus, we still have necessity to share urban video images. Moreover, if we can collect and open a large urban video image dataset, various researchers should come up with and evaluate various image analysis methods.

When sharing urban video images, main obstacle is the privacy concern. Many companies and cities own urban video data, and the data must be useful for opening/sharing to analyze city infrastructure and seasonal change of environment. However, since urban video contains privacy data such as persons and cars, it is difficult to open/share the data widely. In addition, in Europe, the General Data Protection Regulation (GDPR) recently came into force. Therefore, privacy protection mechanism for urban video images should be more important for sharing the video.

There are two main steps to protect privacy of the video. First step is to remove privacy-concern data in the video. This can be achieved by removing video frame which contains privacy data, or anonymizing privacy data such as masking with mosaic pixels. Second step is to secure video sharing process. This can be achieved by using encrypted communication and enhancing traceability. In these steps, this paper focuses the first step—how we can remove privacy data from urban video images. This is because we can reduce privacy invasion risk as much as possible even if the video data could be stolen in video sharing process.

2.3 Anonymization Level

To remove privacy data from the urban video images, we considered the anonymization method by referencing the definition of Chinomi et al. [4]. Chinomi et al. defined that the anonymization of images requires to reduce the concreteness of objects related to privacy. They also defined several anonymization levels such as mosaicking or black out objects which are related to privacy data. In their definition, the abstract degree should be carefully selected, considering the opinion of the user of images and the subjects of images. However, urban images include an unspecified large number of subjects, so that we cannot consider all their opinions. Moreover, in the case of monitoring the city's infrastructure, objects related to privacy such as people and cars are not required. Meanwhile, in their definition, removing (transparence) objects from the images is the most anonymized level. Therefore, based on their definition, our goal in this paper is to make objects related to privacy transparent by removing them naturally.

2.4 Urban Video Anonymization Task

Simple anonymization method cannot be adopted to mobile urban video images. One of the major anonymization methods is utilizing background subtraction [3]. By subtracting the current video frame from the previous video frame, we can acquire an area where something change and where nothing change. Then, the changed area is replaced with the same area from another frame where nothing appears. While the background subtraction method suits for the video from a fixed-point camera, it does not work at mobile urban videos. Although, in the video of a fixed-point camera, the background is fixed, in the video of a mobile camera, the background is changing constantly. The background subtraction method assumes that the background of the input video is fixed. Thus, it is difficult to apply this method to mobile camera videos. In this paper, we process each frame of the video in order to anonymize various videos and images including the ones from mobile cameras.

Additionally, there is another character of the urban videos that the kind, the size, and the position of the object which appears in it might be very diverse. This characteristic makes it difficult to apply the way to detect and remove the target objects such as the pattern/template matching method [14]. Therefore, we have to adopt the method which is flexible to detect and remove various privacy-related objects. After detecting and removing the objects, we have to embed an appropriate image to be natural in the area where the objects have appeared. In summary, we treat the urban video anonymization as the object removal and background completion task in this paper.

3 GANonymizer

In order to tackle the privacy-related object removal and background completion task, we propose an anonymization method, called *GANonymizer*, which consists of two parts of neural networks. In this section, we explain the architecture of GANonymizer and the new layers that we propose to complete the background to be more natural.

3.1 Network Architecture

The architecture of GANonymizer is shown in Fig. 1. In order to detect the target objects from the input image, which might violate the privacy, we adopt the deep

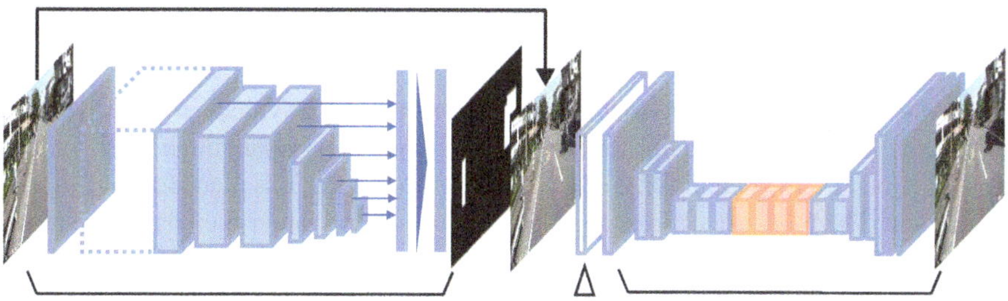

Fig. 1 The architecture of GANonymizer. The target privacy-related objects (e.g., people, car) are detected by bounding boxes via SSD [11]. Then the generated bounding masks are completed by GLCIC [6]

neural networks: Single Shot Multibox Detector (SSD) [11]. SSD is one of the popular models that can detect the object with high accuracy. Especially, we select SSD512 which is the variant SSD model and performs higher than any others. Since the target objects are general and those are contained in PascalVOC dataset [5], we use the model weights which are trained by PascalVOC.

After the target objects are detected, GANonymizer replaces the area where the target objects exist as if there is no object. Although there are a lot of completion methods using computer vision technology, such as PriSurv [4] and PatchMatch [2], the result images are not realistic and are unnatural. On the other hand, the inpainting methods which adopt the deep neural networks [15] succeed in generating the images more realistic and natural. For GANonymizer, we adopt globally and locally consistent image completion (GLCIC) [6] which is one of the most successful models in image completion.

GLCIC is based on generative adversarial networks (GAN) and consists of three networks: the completion network, a local discriminator network, and a global discriminator network. Since GLCIC requires an image and a corresponding binary mask for its input, GANonymizer creates the mask based on the bounding boxes which are the outputs from SSD512. Then GLCIC reconstructs the mask part of the input image based on the whole image and is trained by the procedure of GAN. The local discriminator assesses the quality of the mask part of the image which is completed by the completion network. Simultaneously, the global discriminator assesses the quality of the entire image which is completed by the completion network. The training is terminated when the discriminator networks cannot distinguish between the original input image and the image which is reconstructed by the completion network, that is, the completion network becomes able to reconstruct the mask part of the input image realistically and naturally.

In terms of object removal, it is significant to naturally reconstruct masks based on the various background of images. Hence, for our GLCIC, we apply the model trained with the places dataset, which contains the pictures of the various place, so that it can reconstruct the mask more naturally.

3.2 Auxiliary Context Layer

With the architecture of GANonymizer, most of the urban videos are anonymized precisely. Nevertheless, there are some cases that fail to anonymize well; failure cases are occurred when the size of the mask is too large or it is placed at the edge of the image. The more the objects approach to the camera, larger the size of the mask becomes. It is difficult to reconstruct these large masks since GLCIC reconstruct them by using the contexts surrounding the mask via convolutional operations; there is no information about the center of the large mask for reconstruction. Likewise, the direction of the information of the contexts coming from is restricted because the mask is placed at the edge of the image. In order to solve these problems, we propose the information padding layers, edge shift padding layer, and global feature padding layer that are inserted between SSD512 and GLCIC.

Edge Shift Padding Layer To reconstruct the mask which is placed at the edge of the image, we define the edge shift padding layer (ESP), as shown in Fig. 2(red arrow). ESP copies several pixel rows to the outside of an image. When the edge of the mask is not adjacent to the edge of the image, the pixel rows at the edge of the image are copied. By applying ESP, the input image is extended to the edge side, so that the mask can receive the information from all direction.

Global Feature Padding Layer In order to cope with the large mask, we also propose a global feature padding layer (GFP), which has two steps, as shown in Fig. 2(blue arrow). The first step is to resize the input image into small image so that the size of mask is enough to be able to reconstruct directly and naturally, and

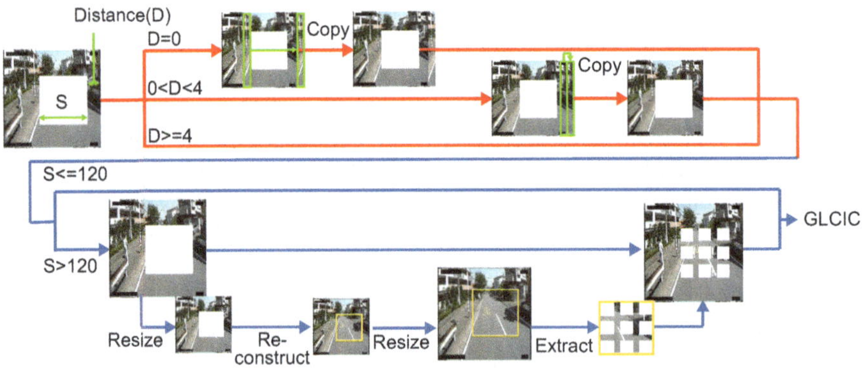

Fig. 2 Auxiliary Context Layer. (Red arrow) This processing temporarily expands input image by copying several pixel rows to the outside of an image, where the several pixel rows are just outside rows of the mask. (Blue arrow) The input image that has large masks is temporarily resized to small and reconstructed. Then, the mask of the input image is pseudo-divided by inserting the first reconstruction result into the original input image with a mesh pattern

then reconstructed. In the next step, the part of a reconstructed mask is extracted as a mesh pattern to use for the information complement. GFP makes it easy for GLCIC to reconstruct the large mask thanks to the mesh pattern.

4 Experiment

In order to validate an effect of GANonymizer, we applied it to actual urban images. Before the experiment, we conducted the preliminary study for ESP and GFP. We show the effect of preprocessing by qualitative evaluation in Sect. 4.1 We then describe the data that we use in our experiment in Sect. 4.2, show the results in Sect. 4.2, and discuss the effect of GANonymizer.

4.1 Preliminary Study

In order to determine some parameters of preprocessing and present its effect, we conducted the preliminary experiments. In the preliminary study, we use a road image extracted from the dataset described in Sect. 4.2 and a pure white image for clearly visualizing the result. The size of both image is 400×400.

Edge Shift Padding In order for ESP to effectively assist the generation of natural images from edge masks, we verify the following two points. First, we look for enough pixels on the edge for the GLCIC to reconstruct the edge mask. Next, based on that, we find the optimal number of pixels for edge padding. In order to verify the above two points, we conducted two experiments. First, we performed GLCIC to the mask whose distance is from 0px to 10px. When the distance is 2px or less, the reconstruction result of the edge mask is not natural due to the lack of surrounding area's information, while reconstruction result is natural when the distance is 3px or more. Then, we performed each ESP in which the padding pixels are random, random pixels in an image, the pixel at the edge of an image, and the pixels on opposite side across the mask. ESP adopting the pixel at the edge of an image was best of four selection, and the second is the pixel on the opposite side. Moreover, because the pasted pixels are not the actual pixels of the image, the width of ESP is better as thin as possible. Therefore, ESP is applied to the mask whose distance is 3px or less and we adopt 4px for the pixel of the edge padding so that we can generate a natural image from the edge mask.

Global Feature Padding We need to examine the size of the mask where GLCIC can directly generate naturally in order that GFP helps GLCIC to generate a natural image from a large mask. In this experiment, We directly reconstruct the large mask whose size is from 50px to 200px at intervals of 10px using GLCIC. Figure 3 is

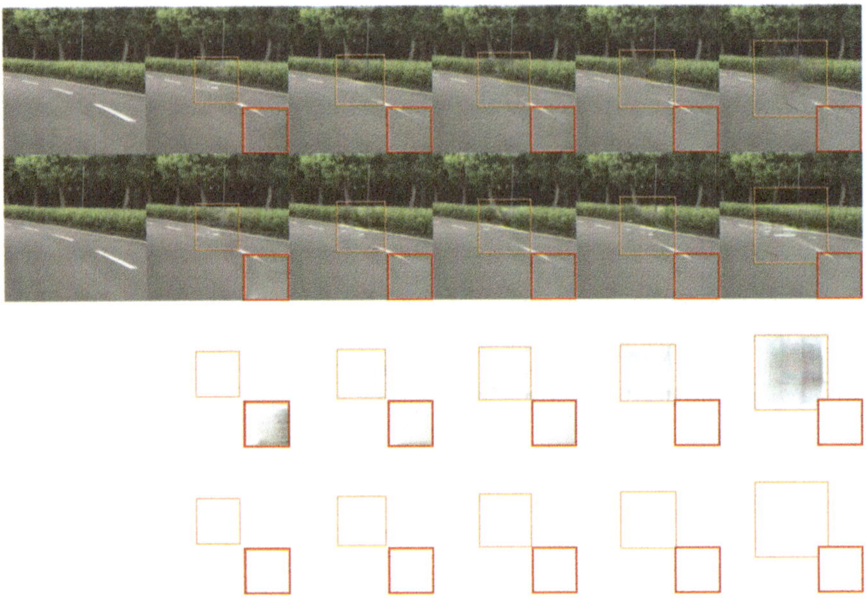

Fig. 3 Preliminary study result. From top row, reconstruction without ESP and with ESP repeatedly. The leftmost images are the original image. Other than that, the image including the large mask (orange) and the edge mask (red). From left to right, the size of the orange mask is {120, 130, 140, 150, 200} and the distance between image edge and red mask edge is {0, 1, 2, 3, 4}

a part of the result of this experiment. The result was enough realistic when the size is 120px or less, while not when the size was 130px or more. Meanwhile, as regards how to divide the mask, when the method that finely divides the mask such as 16 divisions the influence of the lattice between divided small masks is too large. While when the method that coarsely divides such as 4 divisions the influence of the lattice is too small. Therefore, The GFP applied to the mask whose size is 130 px or more and we adopted the 9 divisions for how to divide, so that we could improve the reconstruction of a large mask.

4.2 *Images for Anonymization*

In our experiments, the data is the images taken from an iPhone7 on a car running in the Fujisawa city, Kanagawa, Japan. The total number of frames used in the experiment is 5246, and the size of each image is 1080 × 1920 without resizing. We use the image taken at daytime and evening that is clear and has the risk of privacy invasion. We select persons, cars, buses, bicycles, and bikes as the objects related to privacy, considering the features of the urban images.

Fig. 4 The result of applying GANonymizer to the urban images. Two left images are in daytime and two right images are in evening. The upper row is input images and the lower row is output images

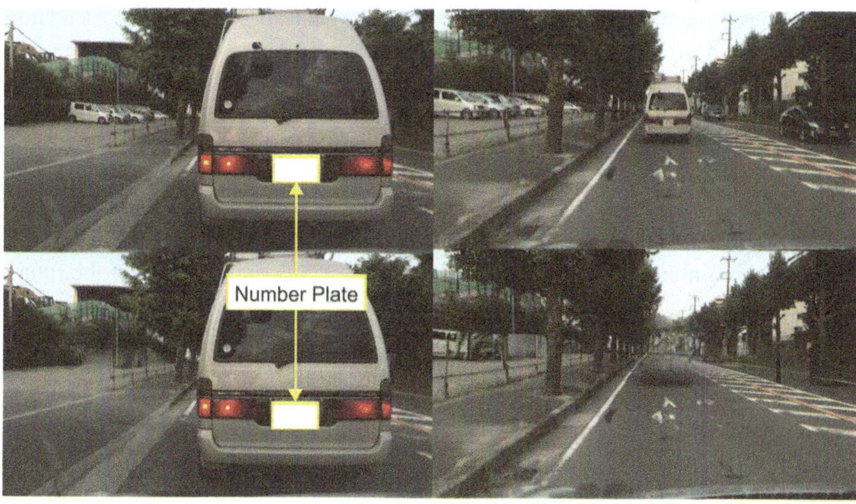

Fig. 5 Failure examples. (Left) The object was too large, which leads missing to detect the object. (Right) The reconstruction of the big car in the center is more blur and coarse than that of the small cars on the left

4.3 Results

Figures 4 and 5 show the results of applying GANonymizer to urban images of the daytime and evening. The results in Fig. 4 show that GANonymizer succeeded to remove persons and cars naturally from the image. The ESP enabled to naturally remove the objects appearing at the edge. Moreover, the GFP improved the problem that the reconstruction of the large mask is not clear. However, as shown in the left of Fig. 5, the detection was difficult in the case that the object appears too large, or the object appears small. Although the GFP improved the problem reconstructing a large mask, the result of removing a large object is still blur and coarse compared to the result of removing small objects (See in Fig. 5). We could remove objects related to privacy, while there are some bad results where some objects are not detected or are not naturally removed. Note that the experiment is conducted on Intel Xeon CPU E 5-1680 v 3 @ 3.20 GHz.

4.4 Discussion

In terms of the privacy-related objects detection, we could precisely detect multi-scale objects related to privacy by using the highest accurate SSD model. However, in the case that a privacy object is extremely close or far, it was difficult to detect objects related to privacy. Figure 5(left) is the example that GANonymizer fails to detect, where the number plate clearly appears. When the object is far, detection failure is not a problem because the object does not clearly appear. On the other hand, when the object is close, privacy problems happen by failing detection. However, detecting a too close object is difficult, especially in the entire figure of the object doesn't appear. There is a possibility that adopting segmentation method can solve this problem. Segmentation categorizes the pixels into the several classes so that it is more likely to partially detect objects than object detection. Therefore, it is likely to improve the detection accuracy by using segmentation instead of object detection or combining them, so integrating it to GANonymizer is one of the future works.

In terms of the object removal, we could reconstruct most masks naturally except for too large masks. Moreover, the two auxiliary context layers help GLCIC to reconstruct difficult masks naturally. Although reconstructing a large mask was improved by the GFP, the reconstruction results of the large mask parts still blur and coarse compared to that of the small mask parts, as shown Fig. 5(right). However, the models based on GAN have the limitation of the size of natural reconstruction. Kingma et al. have proposed Glow that can generate high resolution image [8]. Therefore, it is also the future work to apply Glow image reconstruction in order to naturally reconstruct any size mask.

5 Related Works

Research on the privacy protection of images and videos is divided into two types. The first is the task to challenge the trade-off between privacy protection of training dataset and training performance. The second is the task of challenging the trade-off between the original effectiveness of the image and the privacy protection of the image. The transform is performed in order to reduce the possibility of privacy invasion of the dataset while maintaining the training performance in the specific machine learning task in first type, maintaining the original effectiveness of the data in second type. There are many studies addressing the first task as the growing interest in both privacy protection and machine learning [16, 20]. Meanwhile, our research is classified as the second type because our purpose of application is to both monitor the infrastructure of cities and create the dataset of city image.

When conducting the second type of research, it is necessary to define in advance how much private information should be obscured. Chinomi et al. defined stages of anonymization methods such as blurring and transparency [4]. The definition of that stage is based on the degree to which the method obscures the privacy information. In this definitions, there are many kinds of research on the stage of blurring method such as mosaic to specific parts (human faces and license plates) in the image. Yu et al. have proposed a method to blur people's faces and license plates from images by using anonymization system in which the method that can be mounted on IoT devices with limited resources [21]. There are several studies conducted with similar motivation such as Shu et al. [19]. Although not defined by Chinomi et al., studies on a method of deleting only privacy information from objects related to privacy in images are actively conducted [1]. For example, there is research to replace the face of a person in the image with a generated face that does not actually exist [16]. On the other hand, in this research, the object removal (transparency) which is the highest stage of the anonymization method is regarded as a task to be solved.

The object removal method is basically divided into a phase of object detection and a phase of complementing the detected part. Moreover, these methods differ depending on whether the object is a video or an image. The most rational method for object removal of a video is to process it as just video in each phase since the video can be accessed sequence information. There are many kinds of research that proposed methods of object detection [17] and inpainting [9] for videos. These methods are targeted for the videos from fixed cameras and smartphones, in which the background does not fluctuate greatly, and it is easy to separate foreground and background and follow moving objects of these videos. However, the background of our target video fluctuates greatly because it is taken by a mobile camera whose speed and direction are not constant like a dashcam. For this reason, it is difficult to apply these methods because sequence information cannot be used effectively.

In terms of object removal, Shetty et al. have proposed a deep learning method that learns detection and complementation of specific objects from an image with end-to-end [18]. In this method, it is necessary to input one label of the object to be removed which is existed in the image. However, in urban video images, it is not known beforehand whether objects to be removed exist in a specific frame. Also, since there are several labels to be removed in urban video images, so it is difficult to apply this method to our target videos. Many kinds of research on object detection or inpainting of images have been conducted, but there is no method has been proposed for object removal by integrating the two. Therefore, we propose a method to integrate, by using an auxiliary layer, object detection and image complementation, which removes objects automatically from videos for each frame.

6 Conclusion

In this paper, we presented the novel anonymization method to protect the privacy, called GANonymizer, for the urban videos including mobile cameras. We presented the importance of utilizing the urban images and indicated the problems which occur during anonymizing urban images. In our experiments, we applied GANonymizer to more than 5000 images taken from a smartphone on a car running in the city. As a result, GANonymizer can remove the objects related to privacy, such as people and cars from a single image. Simultaneously, we have found that the information supplemented by the two novel layers helps to reconstruct the large mask and the edge mask. In future work, we plan to remove a significant part of privacy objects such as faces and car number and reconstruct the mask as higher resolution.

Acknowledgement Part of this research was supported by National Institute of Information and Communications Technology.

References

1. Acs, G., Melis, L., Castelluccia, C., De Cristofaro, E.: Differentially private mixture of generative neural networks. In: IEEE Transactions on Knowledge and Data Engineering (2018)
2. Barnes, C., Shechtman, E., Finkelstein, A., Goldman, D.B.: Patchmatch: a randomized correspondence algorithm for structural image editing. ACM Trans. Graph. **28**(3), 24 (2009)
3. Barnich, O., Van Droogenbroeck, M.: Vibe: a universal background subtraction algorithm for video sequences. IEEE Trans. Image Process. **20**(6), 1709–1724 (2011)
4. Chinomi, K., Nitta, N., Ito, Y., Babaguchi, N.: PriSurv: privacy protected video surveillance system using adaptive visual abstraction. In: International Conference on Multimedia Modeling, pp. 144–154. Springer, Berlin (2008)
5. Everingham, M., Van Gool, L., Williams, C.K., Winn, J., Zisserman, A.: The pascal visual object classes (voc) challenge. Int. J. Comput. Vis. **88**(2), 303–338 (2010)
6. Iizuka, S., Simo-Serra, E., Ishikawa, H.: Globally and locally consistent image completion. ACM Trans. Graph. **36**(4), 107 (2017)
7. Kawano, M., Mikami, K., Yokoyama, S., Yonezawa, T., Nakazawa, J.: Road marking blur detection with drive recorder. In: 2017 IEEE International Conference on Big Data (Big Data), pp. 4092–4097. IEEE, Piscataway (2017)
8. Kingma, D.P., Dhariwal, P.: Glow: generative flow with invertible 1x1 convolutions (2018). Preprint arXiv:1807.03039
9. Le, T., Almansa, A., Gousseau, Y., Masnou, S.: Motion-consistent video inpainting. In: ICIP 2017: IEEE International Conference on Image Processing (2017)
10. Liu, Y., Wu, L.: Geological disaster recognition on optical remote sensing images using deep learning. Procedia Comput. Sci. **91**, 566–575 (2016)
11. Liu, W., Anguelov, D., Erhan, D., Szegedy, C., Reed, S., Fu, C.Y., Berg, A.C.: SSD: single shot multibox detector. In: European Conference on Computer Vision, pp. 21–37. Springer, Berlin (2016)
12. Maeda, H., Sekimoto, Y., Seto, T., Kashiyama, T., Omata, H.: Road damage detection using deep neural networks with images captured through a smartphone (2018). Preprint arXiv:1801.09454

13. Morishita, S., Maenaka, S., Nagata, D., Tamai, M., Yasumoto, K., Fukukura, T., Sato, K.: Sakurasensor: quasi-realtime cherry-lined roads detection through participatory video sensing by cars. In: Proceedings of the 2015 ACM International Joint Conference on Pervasive and Ubiquitous Computing, pp. 695–705. ACM, New York (2015)
14. Newson, A., Almansa, A., Fradet, M., Gousseau, Y., Pérez, P.: Video inpainting of complex scenes. SIAM J. Imag. Sci. **7**(4), 1993–2019 (2014)
15. Pathak, D., Krahenbuhl, P., Donahue, J., Darrell, T., Efros, A.A.: Context encoders: feature learning by inpainting. In: Proceedings of the IEEE Conference on Computer Vision and Pattern Recognition, pp. 2536–2544 (2016)
16. Ren, Z., Lee, Y.J., Ryoo, M.S.: Learning to anonymize faces for privacy preserving action detection (2018). Preprint arXiv:1803.11556
17. Shaikh, S.H., Saeed, K., Chaki, N.: Moving object detection using background subtraction. In: Moving Object Detection Using Background Subtraction, pp. 15–23. Springer, Cham (2014)
18. Shetty, R., Fritz, M., Schiele, B.: Adversarial scene editing: automatic object removal from weak supervision (2018). Preprint arXiv:1806.01911
19. Shu, J., Zheng, R., Hui, P.: Cardea: context-aware visual privacy protection from pervasive cameras (2016). Preprint arXiv:1610.00889
20. Wu, Z., Wang, Z., Wang, Z., Jin, H.: Towards privacy-preserving visual recognition via adversarial training: a pilot study (2018). Preprint arXiv:1807.08379
21. Yu, H., Lim, J., Kim, K., Lee, S.B.: Pinto: enabling video privacy for commodity IoT cameras. In: Proceedings of the 2018 ACM SIGSAC Conference on Computer and Communications Security, pp. 1089–1101. ACM, New York (2018)

Performance Evaluation of Source Routing Minimum Cost Forwarding Protocol Over 6TiSCH Applied to the OpenMote-B Platform

Anderson Rocha Ramos, Fernando J. Velez, and Gordana Gardašević

1 Introduction

This work intends to implement SRMCF protocol over 6TiSCH and evaluating its performance by using the OpenMote-B platforms [http://www.openmote.com/]. It is based on the results of a short-term scientific mission (STSM) from COST CA 15104 IRACON during 2017 which took place at University of Banja Luka as a joint activity with researchers from University of Beira Interior (UBI).

A routing protocol is needed to establish end-to-end data delivery. Recent researches have addressed the development of energy-efficient protocols with the aim of finding routes in the networks that minimize energy consumption in nodes with small energy resources [1]. SRMCF is an extension of Minimum Cost Forwarding (MCF) protocols and it is proposed to reduce energy consumption and PLR expending the lifetime of devices in a Wireless Sensor Network (WSN).

The main idea behind MCF is to find the path with minimum cost in a large sensor network [2]. By performing this task, these protocols can compute the effects

This work has been partially supported and funded by COST CA 15104, UID/EEA/500008/2013, CONQUEST (CMU/ECE/0030/2017), and ORCIP. The authors acknowledge the fruitful discussions with Fardin Derogarian and Dragan Vasiljevic.

A. R. Ramos (✉) · F. J. Velez
Instituto das Telecomunicações and Universidade da Beira Interior, DEM, Faculdade de Engenharia, Covilhã, Portugal
e-mail: anderson.ramos@ubi.pt; fjv@ubi.pt

G. Gardašević
Faculty of Electrical Engineering, University of Banja Luka, Banja Luka, Bosnia and Herzegovina
e-mail: gordana.gardasevic@etf.unibl.org

R. José et al. (eds.), *3rd EAI International Conference on IoT in Urban Space*, EAI/Springer Innovations in Communication and Computing,
https://doi.org/10.1007/978-3-030-28925-6_11

of delay, throughput and energy consumption in communications between a node and the sink. The operation of MCF protocols usually comprises two phases [3]. The first phase consists of setting up the cost values to all nodes by broadcasting a message from the sink node, which allows other nodes to adjust their costs according to the received messaged. Next, the source will broadcast the intended message to its neighbors and the nodes that receive the message add their transmission costs. Finally, the node checks the cost in the packet and if the remaining cost is not sufficient to reach the destination, the packet is dropped. Nowadays, WSNs are applied in a variety of scenarios including healthcare, medical monitoring with the so-called wireless body sensor networks (WBSNs) or Medical BANs [1], and urban mobility applications.

Low-power WSNs use a radio transmission technology based on the IEEE.802.15.4-2006 standard, which defines a physical layer and a MAC layer to control the access to wireless medium. The fundamental principle of this standard has in it an intrinsic flaw because it requires wireless devices to keep listening, since they do not know when their neighbors will transmit a message. The IEEE.802.15.4e standard tries to deal with this flaw by altering the MAC protocol while preserving the physical layer [4]. The latter standard implements a schedule that tells the devices in the network when they should transmit, receive, or go to sleep.

Power consumption in sensor nodes is mainly due to the presence of a radio transceiver which plays a decisive role as IoT devices become increasingly more present in everyday applications. Also, recent advances in the production and miniaturization of electronic circuits are causing an increase in the deployment of WSNs.

The OpenWSN project implements the IEEE.802.15.4e Time Slotted Channel Hopping (TSCH) protocol [5] and allows the use of IPv6 communication. This project is considered in the present work to evaluate SRMCF alongside the most recent version of the OpenMote-B platform.

The rest of the paper is structured as follows. Section 2 gives an overview of SRMCF while the OpenWSN stack is analyzed in Sect. 3. Section 4 addresses the process of development and integration of the proposed protocol with the OpenWSN stack. Section 5 describes the OpenMote platforms and parameter settings. Section 6 presents the preliminary results. Finally, Sect. 7 draws conclusions and discusses the possibilities for future work.

2 Source Routing Minimum Cost Forwarding

SRMCF is classified as a reactive protocol, a class of protocols that avoid saving information about network topology in the devices. One example of these protocols is the Gossiping algorithm [6], in which messages forwarded from nodes are sent with some probability to avoid overhead. In general, in reactive protocols there are two types of information traffic. First, there is the traffic between base station (BS) and sensor nodes (SN), when nodes send acquired information to the sink node and receive information from it. Next, there is traffic between sensor node, when these nodes get information data, topology, connections, etc.

In SRMCF protocols, sensor nodes use the MCF protocol to send packets to the base node. However, in this protocol the SN needs to keep information about minimum cost path. The idea of SRMFC is that packets generated at the BS contain information about the path so that SN can use the path information present in the header to route packets, such as in Dynamic Source Routing Protocol (DSR) [7]. The BS is required to have a routing table containing path information that is sent from itself to SN [8] by adding this information to the headers of the packets that are sent in unicast messages.

As a consequence of the SRMCF topology, packets coming from the BS and packets coming from SN have different routing algorithms and also different packet headers. Intermediate nodes are required to identify the appropriate routing scheme based on the type of the message [9]. Data transmission always uses unicast messages. This protocol also supports broadcast messages to perform cost advertisement and cost request. Cost advertisement is generated when BS starts up or when a SN acquires a new cost value, while all SN can broadcast cost request messages [6].

The operation of SRMCF protocol starts with a setup phase, when nodes define their cost to communicate with the BS and the routing table is created. The first part is similar to minimum cost forwarding back-off algorithm, where each node sets its own initial cost as ∞ and the cost to communicate with the BS is equal to 0 [3]. When a node receives a cost value, it compares the received cost with its own cost. If the received value is larger than its own value, the current node value is updated with the new value, which is then broadcasted. This process is executed until all nodes can set up their values to the minimum.

Every time a node needs to change its initial value, in the previous first phase, it sends a packet with its ID and addresses it to a near node. This node adds its own address to the payload and sends this payload to the next node, until the message finally reaches the BS that will now be able to identify all the nodes in the minimum cost path by their IDs and store this information. Figure 1 shows the described setup phase.

The protocol developed during the STSM that inspired the present work is also capable of performing link failure recover. To performer this task, the algorithm keeps a watchdog timer in which each node monitors the activity of its near node.

If a node does not respond during the requested time, its near node begins to broadcast cost request messages so that the network can reestablish the routing information.

3 OpenWSN Protocol Stack

The OpenWSN project is an implementation of the IEEE.802.15.4e standard that allows users to investigate this standard in the context of low-power device-based

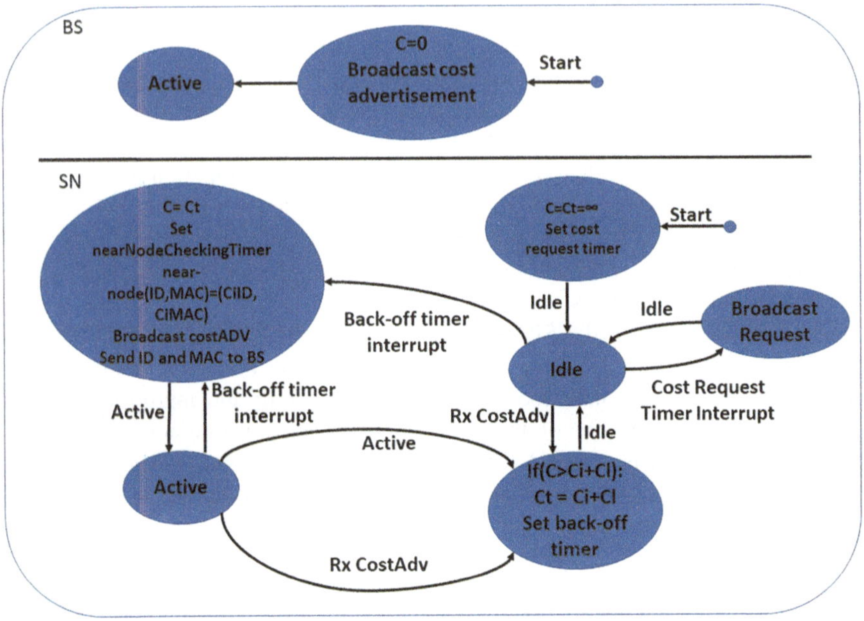

Fig. 1 Network setup phase in SRMCF protocols

networks. It is also heavily based on the concept of IoT, which means that it brings several IoT-based standards, such as Routing Protocol for Low Power and Lossy Networks (RPL), Constraint Application Protocol (CoAP), and IPV6 over Low-Power Wireless Personal Area Network (6LoWPAN). All these standards and the fact that the project can be easily accessed by researchers make the OpenWSN the perfect tool for implementing and testing algorithms over the IEEE.802.15.4e standard. The structure of the OpenWSN stack is illustrated in Fig. 2.

The MAC sub-layer of the stack uses IEEE.802.15.4e TSCH. While this implementation preserves some characteristics of the IEEE.802.15.4 standard, such as Carrier Sense Multiple Access Collision Avoidance (CSMA/CA), which uses the concept of self-interference cancelation to detect collisions [10]. It also brings great improvements, such as the use of channel hopping, which allows persistent multi-path fading and increases interferences immunity [5]. The stack also brings the uRES protocol implemented in it. uRES is used to ensure that nodes can agree in message exchange that occurs in two-way communications since the IEEE.802.15.4e standard by itself has no means to allow efficient data transport over multi-hop paths [5]. This part of the stack may be regarded as a Logical Link Control (LLC) layer. Also, there is a schedule mechanism implemented to allow nodes to identify actions to be taken based on the slots of the frame in which time is sliced in the IEEE.802.15.4e standard.

In the next layer, the stack implements 6LoWPAN as adaptation layer [12]. This layer allows the stack to compress the IPv6 headers. This procedure ensures that

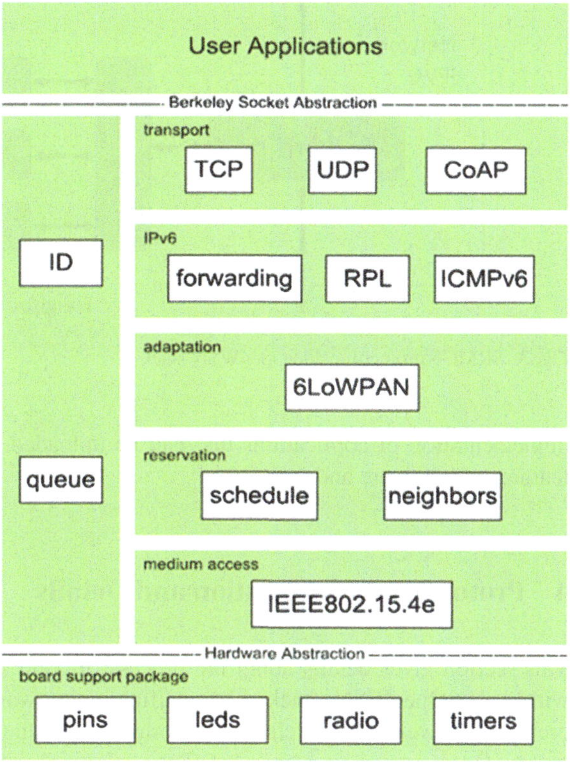

Fig. 2 OpenWSN protocol stack [11]

frames from the IEEE.802.15.4e standard have a length of 127 bytes, at most, and reduces the volume of information required for transmission [11]. Basically, the underling idea is to remove information that is not crucial from the headers while compressing other information, such as source and destination address. The stack also has a mechanism called Low-Power Border Router (LBR) which inflates the headers back to the standard size of the IPv6 headers [13].

RPL is implemented in the next level. It is responsible for the routing topology of the stack. This protocol is designed by the IETF ROLL working group and is especially developed to work with low-power networks. RPL uses the concept of Direct Acyclic Graph (DAG), where a node acts as the root of the topology [14]. Each node in the network receives a rank that keeps the position of that node in the network. These ranks increase in the downward direction. There are two types of messages within this concept. The first is the DAG Information Object (DIO), which is a control message that helps the protocol to build the DAG. The next message type is the Destination Advertisement Object (DAO), which is a unicast control message from nodes to its parents that allows intermediate nodes to recover information about the reverse path of the packet [15].

Finally, the OpenWSN stack implements CoAP [16]. This protocol is constructed as a header which is placed upon the User Datagram Protocol (UDP) and allows the

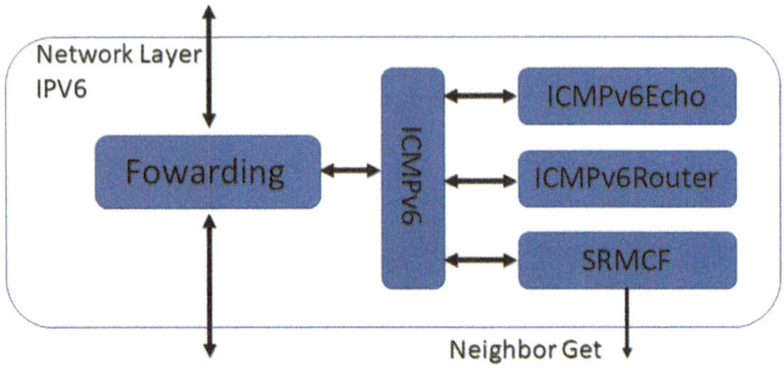

Fig. 3 SRMCF protocol inside OpenWSN stack

implementation of applications that can be uploaded to motes to explore several features of hardware and software.

4 Protocol Implementation and Details

This section gives details about the process of code development and integration within the OpenWSN stack. The starting point was the analysis of the work previously developed, which was performed using the OMNeT++ simulator [17] and the C++ programing language. This code is composed of functions to deal with messages from upper layers, messages from lower layers, messages from SN to BS, messages from BS to SN, self-messages, control messages, and back-off timer.

The first step in this new phase was the development of the described functions in the C programming language. The OpenWSN stack uses the concept of modules. This means that every feature of the stack can be separately worked and then integrated with other layers, what helps to deal with the fact that C is not an object-oriented language. Considering this characteristic, the functions have been implemented in a source file, and a header file was associated to hold prototype functions and variables to be accessed by other modules. Figure 3 illustrates the placement of the SRMCF module inside the stack.

The OpenWSN project is subdivided into firmware and software components [4]. The firmware sub-project contains code that run in the devices and it is where the main part of the implementation process was developed. The original RPL inside this sub-project can deal with message exchanges from BS to SN and from SN to BS in one source code. However, as in SRMCF BS and SN have different routing algorithms; it has been necessary to create two different source codes to be uploaded to each type of device. This fact creates the need for keeping two versions of the firmware, one to hold the routing algorithm to be applied in the SN and another to deal with the BS.

Fig. 4 OpenMote-B devices connected to the Open Visualizer (identified by their MAC addresses)

The process of integration also requires modifications in several source files and modules of the stack. There is a need to create pointers and instances to the new module, and place them in certain modules while identifying functions that are specifically designed for RPL since the goal is to replace the tasks performed by this protocol.

Once the source code is finished and compiled, one way to verify the exchange of message between nodes is using the Open Visualizer (OV). OV is part of the software sub-project and runs in the computer. It has several functionalities, such as the possibility to define a parent node, visualization of traffic between nodes, and identifications of a node by its MAC address. Figure 4 shows the Open Visualizer with two OpenMotes B connected to it. It is possible to observe how the OV identifies the motes by the MAC address from which one can obtain the IPv6 address.

5 Testbed and Settings

OpenMote is an open-source prototyping platform especially designed for the industrial Internet of Things, and it is the latest generation of "Berkeley motes." The OpenMote-CC2538, which is probably the most well-known version of this platform, was, at first, composed of an entire ecosystem where the mote itself was the core of the system and it was followed by other devices, such as a base to allow communication during code development, a USB-based interface designed for ease-of-use by end users and a base to provide power and some basic sensing capabilities through temperature and humidity sensors These motes features a 32 bit ARM Cortex-M3 microcontroller with a radio transceiver of −97 dBm sensitivity level and +7 dBm power transmission [4].

The new versions of the OpenMote platform incorporate most of the above-mentioned features in one single device, what has made code development even more dynamic. These new motes have two radio transceivers that operate at 2.4 GHz and sub-GHz frequency bands. In the Telecommunications laboratory of Instituto de

Fig. 5 OpenMote-B devices from the Instituto de Telecomunicações (IT) of the Universidade da Beira Interior

Telecomunicações/Universidade da Beira Interior, there is a kit of ten units of these motes that are used in experimental research. These motes are shown in Fig. 5.

To evaluate the work and performance of the protocol developed, the code has been uploaded to the motes, with one mote receiving the code related to the BS while being connected to the host computer running Ubuntu 16.04.04 LTS via USB cable.

The first evaluation consists of sending data packets to a mote while waiting for the reply to verify the correct transmission of information. Next, traffic is generated and recorded in the motes' network for different test cases. The traffic of information consists of payloads that are sent by the BS which waits for the reply. To perform the last test, firmware parameters are adjusted to create two scenarios of evaluation, as suggested in [5], as follows:

- Scenario 1: Slot frame length = 20 slots, number of active timeslots in slot frames = 14 slots.
- Scenario 2: Slot frame length = 9 slots, number of active timeslots in slot frames = 5 slots.

These test scenarios allow the measurement of PLR and throughput. The modifications in payload size and time of transmission require the recompilation of the firmware sub-project in each test. As the proposed tests are similar to those performed in previous works found in the literature, using the OpenWSN stack and RPL [5] they allow for the comparison of the data obtained, what justify the chosen scenarios.

OpenWSN implements the IPv6 ping function, which makes possible to send a request to a specific device and observe the time it takes to reply. It is also possible

Table 1 PLR for test scenario 1 and SRMCF

	Payload size		
Hop count	20 bytes (%)	30 bytes (%)	40 bytes (%)
1	0.11	0.10	0.51
2	0.23	0.67	0.70

Table 2 PLR for test scenario 1 and RPL

	Payload size		
Hop count	20 bytes (%)	30 bytes (%)	40 bytes (%)
1	0.13	0.10	0.70
2	0.40	0.85	0.88

Table 3 Throughput [kb/s] for test scenario 1 and SRMCF

	Payload size		
Hop count	20 bytes	30 bytes	40 bytes
1	0.60	0.68	0.82
2	0.65	0.80	0.88

to use a sniffer software, such as Wireshark, to analyze packet content, time of transmission and reply, as well as loss of packets.

6 Preliminary Analysis and Results

The network behavior depends on the number of hops. The results in this section consider a maximum of two hops. Tests with more hops have shown the occurrence of link failures that indicate the need to improve the failure recover mechanism.

The first phase of this analysis has consisted in verifying the correct behavior of the code. To perform this first evaluation, one mote received the algorithm for the BS and another mote received the algorithm for the SN. The BS was connected to the host computer. Next, the IPv6 ping was used to verify if the motes were able to receive packets and send responses. The early stages of code development had indicated that for the stack to work with the new version of the OpenMote B, the baud rate had to be adjusted in some of the modules.

Finally, test scenarios 1 and 2 have been applied. The original RPL of the OpenWSN stack has also been applied in order to evaluate the performance of the two protocols under the considered parameters. Results are shown below.

6.1 Test Scenario 1

This test considers hop distances of 1 and 2 with payload sizes of 20 bytes and 30 bytes, with 300 ms of inter-packet time. Tables 1 and 2 show PLR for the SRMCF, while Tables 3 and 4 show throughput [kb/s] for test scenario 1.

Table 4 Throughput [kb/s] for test scenario 1 and RPL

Hop count	Payload size		
	20 bytes	30 bytes	40 bytes
1	0.50	0.80	0.98
2	0.52	0.75	0.99

Table 5 PLR for test scenario 2 and SRMCF

Hop count	Payload size = 30 bytes (%)
1	0.30
2	0.63

Table 6 PLR for test scenario 2 and RPL

Hop count	Payload size = 30 bytes (%)
1	0.30
2	0.70

Table 7 Throughput [kb/s] for test scenario 2 and SRMCF

Hop count	Payload size = 30 bytes
1	0.40
2	0.49

Table 8 Throughput [kb/s] for test scenario 2 and RPL

Hop count	Payload size = 30 bytes
1	0.50
2	0.50

In scenario 1, it is possible to notice how PLR increases for higher hop counts. One aspect to be considered is how the OpenWSN firmware behaves with different inter-packet times, and how results can be analyzed when inter-packet times are too small. For this test scenario, it can be observed that PLR is smaller for SRMCF in comparison with RPL. Also, higher PLR can be caused by mote queue overflow, since this queue has a limit, in this version of the stack.

6.2 *Test Scenario 2*

This test scenario considers hop distances 1 and 2 with payload sizes of 30 bytes and 300 ms of inter-packet time. Tables 5 and 6 show the PLR, while Tables 7 and 8 show the throughput [kb/s] for the SRMCF and RPL protocols in this test scenario.

The analysis of the results presented in Tables 5 and 6 shows that SRMCF and PRL present the same values of PLR for one hop. However, for distances of two hops SRMCF presents a slightly better performance.

7 Conclusions and Future Work

This work has just presented the analysis and practical results of the implementation of SRMCF in the OpenWSN stack. The performance analysis has shown the potential of the proposed protocol. It has also been possible to observe how the different characteristics of the studied stack may influence communication between motes. It was shown how slot frame length influences network performance, mainly because 6TiSCH schedule algorithm depends on the characteristics of active slots. However, the fact that the stack uses static scheduling makes the influence of characteristics such as slot frame length less significant in network performance. These facts indicate that, to maximize the performance of IoT-based networks while using the OpenWSN stack, one must consider the size of the network and requirements for the transmission time.

As the proposed algorithm brings the potential for more efficient communications among wireless sensor motes, reducing packet collisions and energy consumption, it can enhance data traffic among IoT devices in healthcare, medical monitoring, and urban mobility applications, the deployment scenarios of which are characterized by the existence of interferers and obstacles that tend to increase the packet loss ratio. Besides, OpenWSN OS already implements IPv6, what brings advantages when one considers the context of IoT devices and the constant increasing in the number of devices connected to wireless networks.

The next steps in this research will be the development of functions more suitable to deal with link failure recovery and control data forwarding. Link recovery is a crucial part in SRMCF because once motes in the network have suffered a failure, the network is required to recalculate path costs and send this new information to BSs so that the cost table in it can be updated. The creation of a new graphical interface is also necessary so that one can execute simulations in the graphical environment. Once these features are fully implemented, it will be possible to evaluate the performance of the proposed protocol in comparison to RPL implemented in the OpenWSN stack.

Another feature that can be explored in future work is the development of QoS upon the proposed stack by performing the necessary changes in the algorithm. The team will also make use of the developed protocol to explore long-range communications. The main idea behind this part of the work is to use the OpenMote platform to interoperate with LoraWAN modules that allow long-range communications, e.g., to control swarms of aerial or aquatic drones [18].

References

1. Velez, F.J., Derogarian, F.: Wearable Technologies and Wireless Body Sensor Networks for Healthcare. The Institution of Engineering and Technology, London (2019)
2. Akkaya, K., Younis, M.: A survey on routing protocols for wireless sensor networks. Ad Hoc Netw. **3**(3), 325–349 (2005)

3. Ye, F., Chen, A., Lu, S., Zhang, L.: A scalable solution to minimum cost forwarding in large sensor networks. In: Proceedings of Tenth International Conference on Computer. Communications and Networks (ICCCN), Scottsdale, Arizona, USA, pp. 304–309 (2001)
4. Chang, T., Tuset-Peiro, P., Vilajosana, X., Watteyne, T.: OpenWSN & OpenMote: Demo'ing a complete ecosystem for the industrial internet of things. In: Proceedings of 2016 13th Annual IEEE International Conference on Sensing, Communication, and Networking, (SECON), London, UK (1999)
5. Vasiljević, D., Gardašević, G.: Performance evaluation of OpenWSN operating system on open mote platform for industrial IoT applications. In: Proceedings of 2016 International Symposium on Industrial Electronics (INDEL), Banja Luka, Bosnia Herzegovina, pp. 1–6 (2016)
6. Patra, C., Botezatu, N.: Effect of gossiping on some basic wireless sensor network protocols. In: Proceedings of 21st International Conference on System Theory, Control and Computing (ICSTCC), Sinaia, Romania (2017)
7. Ahmad, S., Awan, I., Waqqas, A., Ahmad, B.: Performance analysis of DSR & extended DSR protocols. In: Proceedings of 2nd Asia International Conference on Modelling and Simulation (AMS), Kuala Lumpur, Malaysia, pp. 191–196 (2008)
8. Derogarian, F.: Design of a body sensor network embedded in textiles for biomedical applications. Doctoral Dissertation, University of Porto, Porto, Portugal (2015)
9. Derogarian, F., Ferreira, J.C., Tavares, V.M.G.: Analysis and evaluation of an energy-efficient routing protocol for WSNs combining source routing and minimum cost forwarding. J. Mob. Mult. **14**(4), 469–504 (2018)
10. Garcia-Luna-Aceves, J.J.: Carrier-sense multiple access with collision avoidance and detection. In: Proceedings of 20th ACM International Conference on Modelling, Analysis and Simulation of Wireless and Mobile Systems, Miami, Florida, USA, pp. 53–61 (2017)
11. Watteyne, T., et al.: OpenWSN: a standards-based low-power wireless development environment. Eur. Trans. Telecommun. **23**(5), 480–493 (2012)
12. Melorose, J., Perroy, R., Careas, S.: Compression format for IPv6 datagrams in low power and lossy networks. Statew. Agric. Land Use Baseline. **1**, 1–25 (2015)
13. Winter, T., et al.: RPL: IPv6 routing protocol for low power and lossy networks (work progress). http://tools.ietf.org/html/draft-ietf-roll-rpl-19, pp. 1–164 (2011)
14. Airehrour, D., Gutierrez, J.A., Ray, S.K.: A Trust-Aware RPL routing protocol to detect blackhole and selective forwarding attacks. Aust. J. Telecommun. Digit. Econ. **5**(1), 50 (2017)
15. Wang, D., Tao, Z., Zhang, J., Abouzeid, A.A.: RPL based routing for advanced metering infrastructure in smart grid. In: Proceedings of 2010 IEEE International Conference on Communications Workshops, Capetown, South Africa (2010)
16. Shelby, Z., Hartke, K., Bormann, C., Frank, B.: "Constrained application protocol (CoAP)", draft-ietf-core-coap-18 (work in progress) (online). http://tools.ietf.org/html/draft-ietf-core-coap-18 (2013)
17. Köpke, A., et al.: Simulating wireless and mobile networks in OMNeT++ the MiXiM vision. In: Proceedings of First International ICST Conference on Simulation Tools and Techniques for Communications Networks and Systems, Marseille, France, pp. 71–78 (2008)
18. Velez, F.J., et al.: Wireless sensor and networking technologies for swarms of aquatic surface drones. In: Proceedings of IEEE 82nd Vehicular Technology Conference (VTC2015-Fall), Boston, MA, USA (2015)

Part IV
Urban Mobility and Connected Vehicles

Analysis of Time of Use and Intermodality of Ride-Hailing Services in Singapore Using Mobile Web Traffic Data

Sau Yee Chan

1 Introduction

App-based ride-hailing services are increasingly popular in Asian cities. In Singapore, for instance, there are now 1.5 times more vehicles registered to provide ride-hailing services than taxis [1]. Compared to taxis, ride-hailing services are shown to lead to reduced or more consistent wait times and increased utilization rate [2, 3]. From an economic standpoint, ride-hailing services meet a demand that was previously unsatisfied by the taxi industry, by matching demand and supply in real-time and lowering transactional costs [4]. Pricing strategies such as surge pricing also offer more financial incentive to drivers and encourage a higher supply of vehicles. This increase (including *perceived* increase) in both demand and supply thus leads to a higher probability of getting matched [5]. These qualities render ride-hailing a potential rival to public transport.

Efforts to quantify the impact of ride-hailing services on public transport are hindered by the high cost (in terms of both money and time) of large-scale household surveys [6] as well as the unavailability of ride-hailing trip data [2]. As a result, related literature often makes use of ad-hoc surveys, which suffers from the issue of representativeness and potential bias.

To the best of the author's knowledge, this is the first study to propose the use of mobile web traffic data to assess ride-hailing usage. The dataset contains the web

Contribution during Master's programme at Ecole des Ponts ParisTech (France) and an internship at Teralytics Pte Ltd (Singapore).

S. Y. Chan (✉)
AGOOP Corp., Tokyo, Japan
e-mail: sauyee.chan@agoop.co.jp

R. José et al. (eds.), *3rd EAI International Conference on IoT in Urban Space*, EAI/Springer Innovations in Communication and Computing,
https://doi.org/10.1007/978-3-030-28925-6_12

browsing records of two million anonymized mobile phone subscribers of a telecom company in Singapore that has a market share of about 20%. In addition to the general advantages of mobile phone data such as its timeliness, size of samples, and high spatial-temporal resolution [7], mobile web traffic data is uniquely suitable for ride-hailing analysis in that ride-hailing users must connect to the Internet to request for rides on the platform. It is therefore possible to identify the ride-hailing app used and to estimate the start and end time of each trip, based on relevant web browsing attributes.

This study makes use of real-world mobile web traffic data on one weekday to analyze the use of ride-hailing services. A mobile penetration rate of about 150% [8] makes Singapore an appropriate choice of study using this methodology. In light of Singapore's existing hub-and-spoke transport network where almost 20% of working adults travel to work by combining the train with another mode [9], we focus on the intermodality of trips from home to work by examining the spatial and temporal patterns of ride-hailing in two residential neighborhoods.

2 Related Work

Impact of Ride-Hailing on Public Transport Studies in the USA present conflicting findings on ride-hailing's impact on public transport while independent evidence is scarce in Europe. The American Public Transportation Association's study in seven American cities suggests that ride-hailing complements public transport by serving users in off-peak hours, when headway is the longest [10]. In contrast, some studies [11, 12] report that ride-hailing substitutes public transport in American cities.

In addition to time of use (*when*), the question of *how*, including intermodality of ride-hailing services, is of great interest. The potential of ride-hailing services in facilitating first and last-mile public transport connections has been discussed [11]. In Europe, a study commissioned by Uber on the impact of Uber services on users' mobility in France and Switzerland found that 12–47% of TNC trips connect to public transport before or after the ride [13].

To the author's knowledge, studies in Asian cities have not been published. It is hence unclear how ride-hailing services are used in relation to public transport in these cities where population density and mode share of public transport are generally higher than that of their American or European counterparts.

Use of Mobile Data in Transportation Planning Data captured by telecom companies through users' mobile phones have seen increasing application in transportation planning. Compared to census' usual sampling rate of 1–2%, mobile phone data can cover 20–30% of the population depending on the telecom company's market share and mobile penetration rate. Its accuracy is said to be comparable with register-based census data and it has the advantage of being independent of subjective responses [14].

Another oft-used form of mobile phone data is the call detail records (CDR), which contain the timestamp and location of phone calls and sometimes that of messages [15]. CDR data have been used to estimate travel demand [15], characterize land use [16], and infer human mobility patterns [17]. Unlike the CDR, mobile web traffic data provide information about a subscriber's web browsing history, including the domain name and other properties such as download and upload volumes. It is hence highly relevant for research on ride-hailing services, which are online-enabled platforms connecting passengers with drivers.

Public Transport Network in Singapore Public transport forms the backbone of Singapore. According to the latest Household Interview Travel Survey [18], public transport accounts for 50% of all motorized trips and 63% of all motorized trips during peak hours. Car ownership is low due to government policies, including but not limited to the vehicle quota system (VQS) [19].

Public transport in Singapore follows a hierarchical system. Singapore's main modes of transport comprise the mass rapid transit (MRT), the light rail transit (LRT), and buses. LRT and buses act as feeder services to connect areas in housing estates to MRT stations, while MRT serves long-haul travel [20].

3 Methods

Ride-hailing trips and MRT trips are detected using mobile web traffic data. Anonymized web traffic records of mobile phone users on a weekday are provided by a telecom company in Singapore with two million subscribers. Each record contains an anonymized user ID, a cell ID, a timestamp (in milliseconds), domain name, upload volume, and download volume. Each mobile phone user generates an average of 4–5 events per minute, when he or she uses the phone to upload or download content from the Internet.

We make use of a cell tower database of the 3G and 4G networks provided by the telecom company to locate the area in which the web traffic has occurred. For simplicity, it is assumed that a mobile phone user's location in each record is equivalent to the location of the cell tower.[1] No triangulation is performed.

Aim of Analysis and Algorithm Our aim, firstly, is to examine the time of use of ride-hailing services in Singapore. Secondly, it is to explore the extent of intermodality of ride-hailing services with public transport. To this end, after the necessary preprocessing, we carry out ride-hailing trip detection followed by MRT (i.e., train) trip detection. Each trip taken T can therefore be defined as follows:

$$T = \{(u, t, m, p_{\text{coords}}, p_{\text{str}})\}$$

[1]With over 22,000 cells covering the city-state, the average area of each cell is 33,000 m^2 (equivalent to a 180 m by 180 m mesh).

where u is the anonymized user ID, t the start time, m the mode of transportation (ride-hailing or MRT), p_{coords} the estimated start position in latitude and longitude, and p_{str} the human-readable name of the estimated start position.

Noise Removal To ensure that the indicated cell ID is a reasonable estimation of the location of the mobile phone user, we filter out transient hand-offs and increase robustness by performing an extra aggregation step. If the user generated n events across different cells in one minute, we consider his/her location to be the mean position from the positions of all n events.

Ride-Hailing Trip Detection Given the specific domain names used by ride-hailing apps in Singapore (namely Uber and Grab, at the time of analysis) and the observation that upload and download volumes display specific patterns at the time of trip start, we define the start time and start location of ride-hailing trips.

Train Trip Detection A mobile phone user's presence can be detected at underground MRT stations to great accuracy, as underground train stations require dedicated equipment and therefore have specific cell IDs. To ensure accuracy, the second part of our analysis is performed on two residential neighborhoods with train stations on a train line that runs entirely underground. Both neighborhoods are new towns located at about 10 km north-east of the Central Business District (CBD) of Singapore.

4 Results

Ride-Hailing Use by Time of Day Ride-hailing use coincides with peak hours on public transport, on both the national level and the local level (Fig. 1). Out of approximately 100,000 rides detected, two sharp peaks can be observed in ride-hailing trip starts. They correspond to the morning peak at 7–9 a.m. and the evening peak at 6–8 p.m.. This suggests that, unlike in the USA or in Europe, ride-hailing in Singapore is used when demand for public transport is at the highest.

Intermodal Ride-Hailing: MRT Trips There are few intermodal trips involving ride-hailing and subsequently the train. Only about 5–6% of ride-hailing trips starting in both neighborhoods connect to a MRT station within an hour (Table 1). Out of these trips that connect to a train station within an hour, 40–60% connect to a station along the same train line instead of the closest station. Hence, even when trips are intermodal, ride-hailing users' behavior prove to be strategic, opting to connect to a train station that is not necessarily the closest. One possible explanation is that TNC users, who tend to be affluent and technology-savvy [2], try to utilize TNC services as a way to reduce interline transfer or to decrease travel time.

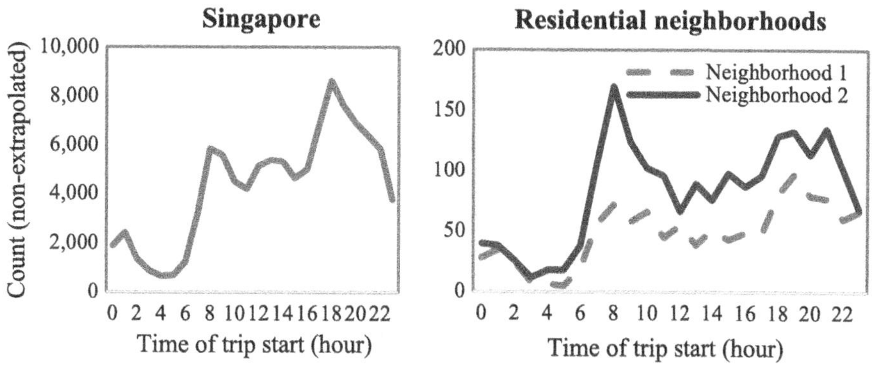

Fig. 1 Ride-hailing trip starts by hour in Singapore (left) and in two residential neighborhoods (right). Note numbers are not extrapolated

Table 1 Train stations connected to after ride-hailing trip

		MRT trips preceded by ride-hailing trips			
		Neighborhood 1 ($n = 1187$)	Percentage (%)	Neighborhood 2 ($n = 1995$)	Percentage (%)
Train station breakdown	Nearest train station	25	37	16	18
	Station *on the same line*, excluding the nearest train station	26	39	57	63
	Stations *on other train lines*	16	24	17	19
Total		67	100	90	100

5 Discussion

Anonymized web traffic records of two million mobile phone subscribers in Singapore have been used to analyze time of use and intermodality of ride-hailing trips. While this study is far from extensive, it is probably the first to propose the use of mobile web traffic data in analyzing the nature of ride-hailing services and serves as a proof of concept that mobile web traffic data provide interesting insights into when and how ride-hailing services are used.

The findings demonstrate that, first, ride-hailing usage in a dense Asian city, namely Singapore differs from the patterns seen in European or North American cities. Second, ride-hailing trips appear to be point-to-point trips instead of intermodal ones that integrate with existing transport infrastructure.

This study sheds light on how ride-hailing services could challenge the traditional hub-and-spoke transport model conceived by the government in the long run. As seen in the case of Singapore, the impact of ride-hailing is not limited to modes serving shorter distances such as the feeder buses, the LRT, or taxis. Ride-hailing is seldom used in conjunction with the train and hence competes with it. Even though ride-hailing currently only represents a small fraction of all motorized trips (when extrapolated), this implies greater competition for public transport in general, as well as greater impact on the city level in terms of vehicle miles traveled and congestion in the CBD.

For further analysis, we can consider extending the scope of study to multiple days, weeks, and neighborhoods. Other contextual information such as the socio-economic status of users, trip purpose, travel time, and cost would complement the dataset and help planners understand the underlying factors that influence mode choice.

References

1. Tan, C.: Private-hire cars outnumber taxis by a mile. The Straits Times, Singapore, 24 May 2017
2. Rayle, L., Dai, D., Chan, N., Cervero, R., Shaheen, S.: Just a better taxi? A survey-based comparison of taxis, transit, and ridesourcing services in San Francisco. Transp. Policy. **45**, 169 (2016)
3. He, F., Shen, Z.J.M.: Modeling taxi services with smartphone-based e-hailing applications. Transp. Res. C Emerg. Technol. **58**, 93–106 (2015)
4. Aguiléra, A., Rallet, A.: Mobilité connectée et changements dans les pratiques de déplacement. Réseaux. **200**(6), 17 (2016)
5. Harding, S., Kandlikar, M., Gulati, S.: Taxi apps, regulation, and the market for taxi journeys. Transp. Res. A. **88**, 15–25 (2016)
6. Ge, Q., Fukuda, D.: Updating origin-destination matrices with aggregated data of GPS traces. Transp. Res. C. **69**, 291–312 (2016)
7. Lenormand, M., Ramasco, J.J.: Towards a better understanding of cities using mobility data. Built Environ. **42**(3), 356 (2016)
8. Infocomm Media Development Authority: Mobile penetration rate, Singapore (2017)
9. Department of Statistics: Census of population 2010 statistical release 3 geographic distribution and transport, Singapore (2010)
10. American Public Transportation Association: Shared Mobility and the Transformation of Public Transit. Transportation Research Board, Washington, DC (2016)
11. Shaheen, S., Chan, N.: Mobility and the sharing economy: potential to facilitate the first-and last-mile public transit connections. Built Environ. **42**(4), 573–588 (2016)
12. Babar, Y., Burtch, G.: Examining the impact of ridehailing services on public transit use. SSRN Electron. J., 1–54 (2017). https://doi.org/10.2139/ssrn.3042805
13. 6t-Bureau de Recherche: Uses, users and impacts of private hire services: survey of users of the Uber application (2015)
14. De Meersman, F., et al.: European Conference on Quality in Official Statistics, pp. 1–16 (2016)
15. Toole, J.L., Colak, S., Sturt, B., Alexander, L.P., Evsukoff, A., Gonzalez, M.C.: The path most traveled: travel demand estimation using big data resources. Transp. Res. C. **58**, 162–177 (2015)

16. Soto, V., Frias-Martinez, E.: Robust land use characterization of urban landscapes using cell phone data. In: Proceedings of the 1st Workshop on Pervasive Urban Applications, in conjunction with 9th International Conference on Pervasive Computing (2011)
17. Jiang, S., Ferreira, J., Gonzalez, M.C.: Activity-based human mobility patterns inferred from mobile phone data: a case study of Singapore. IEEE Trans. Big Data. **3**(2), 208–219 (2017)
18. Land Transport Authority: Household interview travel survey 2012, Singapore (2012)
19. Han, S.S.: Managing motorization in sustainable transport planning: the Singapore experience. J. Transp. Geogr. **18**(2), 314–321 (2010)
20. Haque, M.M., Chin, H.C., Debnath, A.K.: Sustainable, safe, smart-three key elements of Singapore's evolving transport policies. Transp. Policy. **27**, 20–31 (2013)

An Exploratory Study of Relations Between Site Features and I2V Link Performance

Luís Sousa, Pedro M. Santos, and Ana Aguiar

1 Introduction

Smart cities will require new solutions for connectivity and data collection inspired by the Internet-of-Things (IoT) paradigm [2]. Opportunistic collection of data produced by road-side IoT nodes by wireless-enabled vehicles is one of such strategies. When planning the deployment of such road-side nodes, an important challenge is to evaluate if the node will be able to transfer all of its data to passing vehicles. Learning this requires resource-consuming measurement campaigns at several potential locations, or accurate estimation methods of I2V transfers that rely on models of throughput vs. distance and mobility traces.

An alternative approach is to estimate data transfer rates and volume from relevant qualitative characteristics of a potential deployment location. Given the vehicular nature of the receivers, such features are typically related to mobility: these may be directly related to the vehicular mobility patterns (e.g., speed distribution and stopped/moving periods of the vehicles), or indirectly by mobility-affecting points-of-interest (POIs), such as traffic (e.g., traffic lights, crosswalks) and/or fleet-related infrastructure (e.g., bus stops, garbage bins). In this paper, we identify mobility and throughput patterns that can be associated with such features, leveraging a dataset of Wi-Fi link characterization in I2V connections. We observe that a large percentage of connection samples take place when the vehicles are stopped, and relate stopping areas to nearby POI.

L. Sousa · P. M. Santos (✉) · A. Aguiar
Faculdade de Engenharia, Universidade do Porto, Porto, Portugal
Instituto de Telecomunicações, Lisbon, Portugal
e-mail: lm.sousa@fe.up.pt; pmsantos@fe.up.pt; anaa@fe.up.pt

R. José et al. (eds.), *3rd EAI International Conference on IoT in Urban Space*, EAI/Springer Innovations in Communication and Computing,
https://doi.org/10.1007/978-3-030-28925-6_13

Our main contributions are:

- Identification of mobility and throughput patterns associated to I2V service from analysis of a real-world link quality dataset; and
- Elation of patterns and POI of a site and break down per POI class.

The remainder of this article is organized as follows. In Sect. 2 we review the literature on this topic. We describe the I2V dataset and experimental setup, and uncovered mobility and throughput patterns in Sect. 3. In Sect. 4, we relate identified mobility and throughput patterns with POIs of each site. Final remarks are drawn in Sect. 5.

2 Related Work

Vehicular data collection from infrastructure nodes has been proposed in [6]. The authors report a distributed mobile sensor system: data collected by sensors installed in vehicles is offloaded to static road-side units. The work of [3] builds on the previous by allowing vehicles to receive data from the static nodes. In previous work [11], we characterized the transfer rates, volumes, and connection times in an I2V scenario—a traffic light-bound wireless sensor that transferred data to passing buses equipped with on-board access points (APs). This work was smaller in scale (only one site) and the impact of nearby traffic and fleet-related elements was not studied in detail; the focus was put on relating observed performance with bus schedules (an approach that is not possible in this articles as the measurements reported involved garbage trucks). As we also demonstrated in [12], the opportunistic nature of V2X contacts lends itself to the use of delay-tolerant protocols (DTN).

There is numerous works on IEEE 802.11 link characterization and data volume measurements in vehicle-to-X scenarios. The works reported in [10] and [14] evaluate the performance of IEEE 802.11-based I2V links to provide Internet to vehicular users in highway scenarios (in [14], also car-to-car measurements are taken). The authors of [5] describe range, association times, UDP, and total data volume between infrastructure and cars at different speeds. In [1], the performance of IEEE 802.11 is characterized under aggressive mobility scenarios (node mobility up to 240 km/h). The tools to perform bandwidth estimation include the well-established Iperf [9] application, that attempts to send the maximum data possible through the channel to estimate the channel capacity. WBest [7] offers an alternative for end-to-end capacity measurements: it assumes that end-to-end capacity is defined by the narrowest link, that "the time dispersion between the two packets is linearly related to that narrow link capacity" [7]. To the best of our knowledge, this is the first work proposing the estimation of I2V data transfers from site features.

3 Experimental Setup and Dataset Analysis

We use I2V measurements acquired in the context of the PortoLivingLab platform [13]. PortoLivingLab is a smart city-enabler IoT platform deployed in Porto, Portugal, comprising sub-platforms *UrbanSense* [8], a collection of 20 sensing units equipped with a Wi-Fi module and named Data Collection Units (DCUs), and *BusNet*, a vehicular network of 600+ on-board units (OBUs) installed in the public transportation and waste disposal fleets of Porto (the former accounting for 400 nodes) and that offer a Wi-Fi hotspot service. An initial characterization of I2V Wi-Fi links in the PortoLivingLab setup is reported in [11].

The I2V experimental dataset was collected at three DCUs, deployed at three disparate sites referred to as “A”, “B”, and “C”, from measurement sessions carried out with the OBUs installed in the waste disposal fleet. The DCUs, configured as Wi-Fi clients, connected opportunistically to passing OBUs (configured as access points, AP) to collect the GPS of the vehicles and perform measurement sessions of Iperf and WBest. The resulting dataset contains collections of samples (i.e., measurement tuples) composed of timestamp, MAC address of OBU, position and speed of vehicular node, and link quality metrics—throughput, PLR, jitter—from both tools. The three steps—collecting GPS and running Iperf and WBest measurement sessions—were performed sequentially and repeated in this order while the connection lasted. The time between tuples is at least 2 s, as the GPS data of the vehicle is obtained via a SSH-based query to the OBU, and the IPerf measurement session is scheduled to last 1 s. Individual connections to OBUs are identified by aggregating samples that are temporally close (less than 60 s apart). The data volume of a connection is the summation, over all samples, of the product of each sample throughput and the time interval until the next sample. In pre-processing, samples lacking valid GPS data were discarded. In total, we obtained 12,369 link quality samples, 588 connections, and detected 16 different OBUs.

3.1 Evaluation of Bandwidth Estimation Tools

As mentioned earlier, the available dataset contains channel capacity measurements from two tools that operate in significantly different ways. Iperf injects the maximum possible data in the channel, and it can be highly disruptive to other on-going connections. WBest proposes a non-disruptive alternative, and thus it may be better suited when one of the terminals is engaged in other unrelated connections.

We seek to understand how similar the estimates of both tools are. In the I2V dataset, the samples for each tool were obtained sequentially: the measurement session by Iperf, lasting 1 s, is followed by a session by WBest. In Fig. 1, we plot the relation of the Iperf and WBest values for the same measurement tuple, for the cases of stopped vehicles (speeds under 1 km/h), slowly moving (speed between 1 and 6 km/h), and fast speed (over 6 km/h). Linear regression parameters and standard

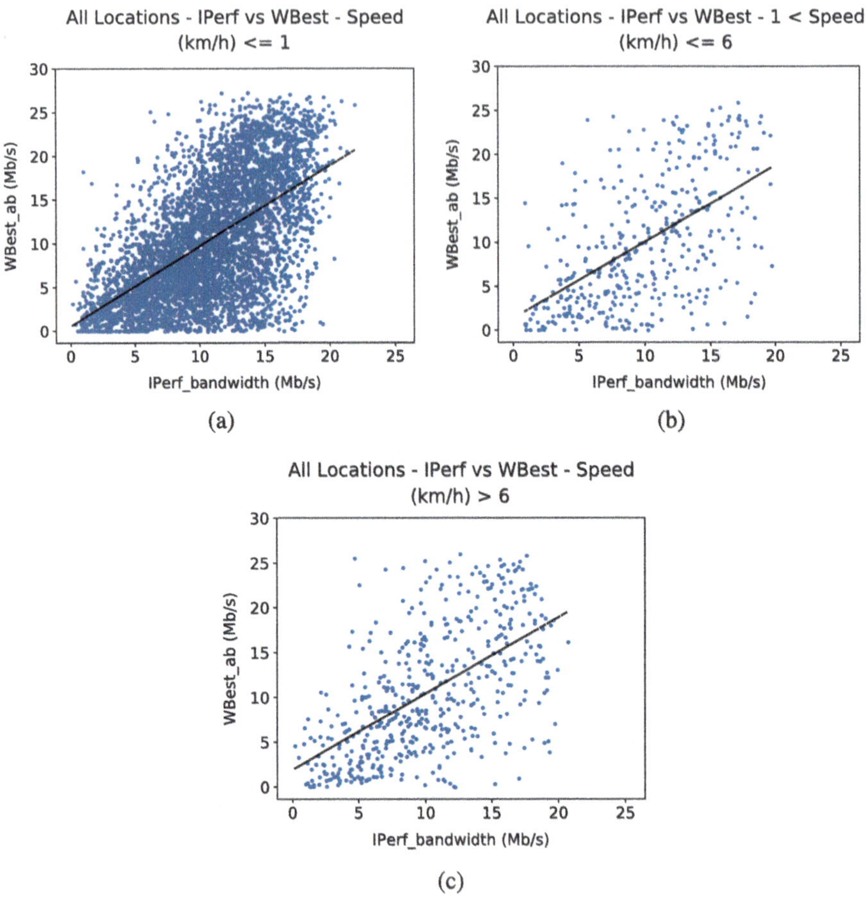

Fig. 1 Comparison of Iperf/WBest measured throughput, for same measurement tuple. (**a**) Slow moving vehicles. (**b**) Medium speed. (**c**) Fast moving vehicles

Table 1 Linear regression for Iperf/WBest data

Case	Slope (m)	Y-intersect (b)	Std. dev. of res.	Pearson corr.
Slow	0.92	0.52	5.47	0.59
Medium	0.87	1.37	5.94	0.56
Fast	0.85	1.90	5.73	0.58

deviation of residuals are also shown in Table 1. Observed slopes are close to 1 for most cases, particularly for slow moving vehicles. The slope decreases as the speed increases, indicating that WBest under-estimates the bandwidth, which in turn may be a consequence of increased packet loss. For the remainder of this paper, we use the Iperf measurements.

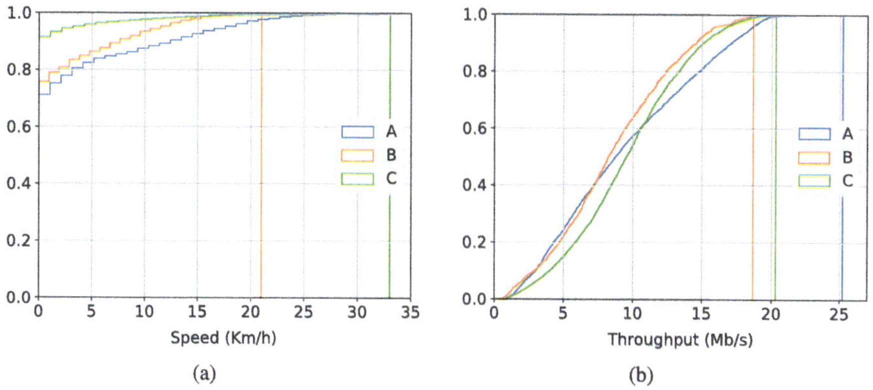

Fig. 2 CDF of relevant metrics for all sites. (**a**) Speed samples. (**b**) Throughput samples

3.2 Analysis of Vehicle Mobility and I2V Throughput

We analyzed the collected data by evaluating how the measured link throughput varies with several mobility-related features, particularly the vehicles' speed and distance to the DCU. The speed of the vehicles during the I2V measurements is shown in Fig. 2a. We observe that, across the three sites, the large majority of samples was taken when the vehicle is stopped. The ratio of stopped and moving intervals of the vehicles was, on average, 83.32%, so there are more throughput samples collected when the vehicles are stopped.

We also observed the throughput samples to follow similar distributions at the three sites, as can be seen from the throughput CDFs in Fig. 2b. The performance of throughput versus distance is depicted in Fig. 3 for 10 meter-wide bins. We observe that the communication range differs from location to location, which can explained by the different road topologies, and that there are distance intervals where a larger number of samples occur (see top axis of graphs), indicating that at those distances there may be points-of-interest (or areas affected by nearby POIs).

The two main takeaways are that: (a) in all locations, most of the connection samples occur when vehicles are stopped; (b) sample frequency w.r.t. distance varies between distance intervals and from location to location. These two observations lead us to conclude that, at each location, POIs at particular distances create stopping opportunities.

4 Identification and I2V Performance at POI

We hypothesize, from the previous analysis, that most data transfers occur at locations where points-of-interest exist. We define two classes of POIs:

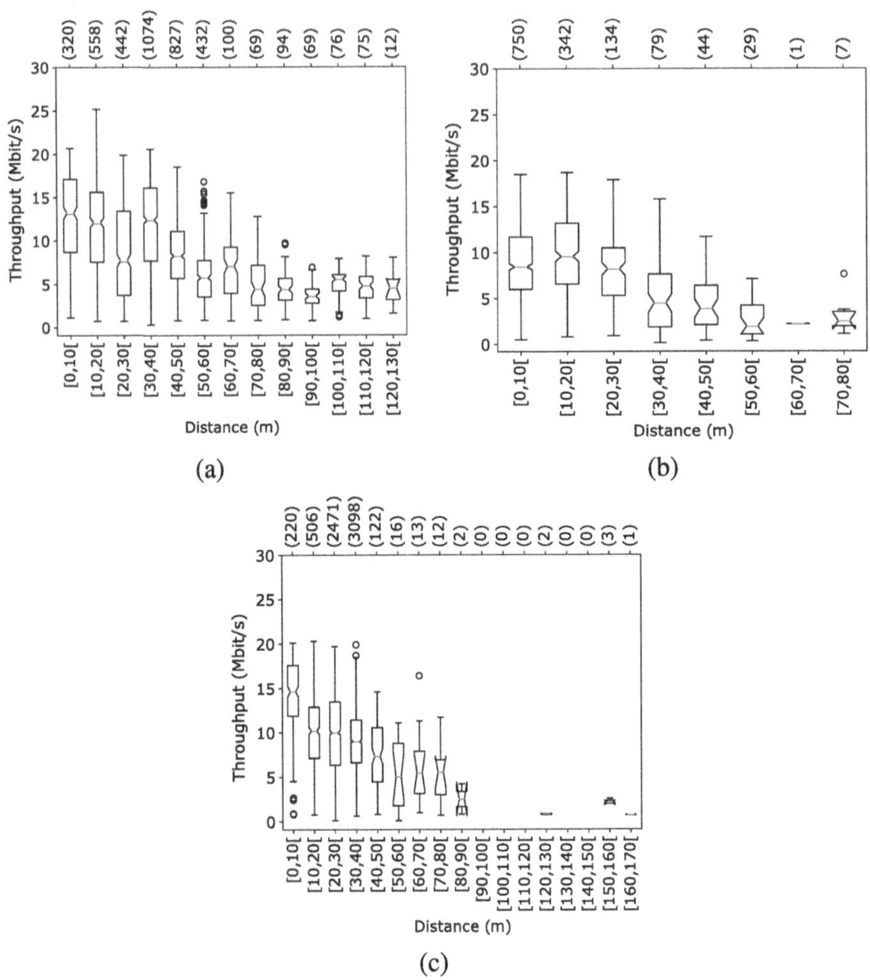

Fig. 3 Throughput vs. distance at the selected locations (top axis: number of samples). (**a**) Site A. (**b**) Site B. (**c**) Site C

- traffic-related, e.g., traffic lights, crosswalks;
- fleet-related, e.g., bus stop, garbage bins.

4.1 Identifying Stopping Regions and Association to POI

We further analyzed the I2V measurement samples at which vehicles were stopped. We consider a vehicle is "stopped" if its speed is inferior to 3 km/h. We verified by visual inspection that these samples were not uniformly distributed over the roadways, and that their spatial density is higher near POIs. We set out to confirm the link between POIs and "stopped" samples as follows.

In a first stage, we applied a density-based clustering technique, DBSCAN [4], to identify areas with high concentration of samples. The DBSCAN parameters were: neighborhood radius—2.5 m; minimum number of points to become a core point—8. The original dataset was sub-sampled prior to the cluster algorithm, to retain the highest-throughput samples. We defined consecutive 10 m-wide rings centered at the DCU location, and take only the 40% highest-throughput samples in each ring. This approach allows us to sub-sample homogeneously over the entire range of distances. The parameters of the sub-sampling and DBSCAN procedures were found after iterative search: at each round, we evaluated empirically the number and size of produced clusters. The goal was to obtain a manageable number of clusters for the subsequent POI assignment task (as opposed to having many small clusters or few large ones).

After the clustering algorithm was applied, we established an univocal correspondence between clusters and POIs. We identified manually, on each site, the following types of POIs: traffic lights, crosswalks, and garbage bins. We computed the Euclidean distance between cluster centroids and POI locations, and associated the closest ones. We concluded that this approach did not perform well in some cases. E.g., oftentimes a slender cluster, known to be caused by a particular traffic light, presents its centroid closer to the crosswalk of an inflowing parallel street. Thus, we associated manually POIs to clusters, taking into consideration: (a) whether POI is inside the cluster; (b) direction of traffic flows. POIs may be assigned more than one cluster.

Table 2 presents a quantitative characterization of the datasets corresponding to each of these steps—the initial dataset, and those associated with the produced clusters and assigned clusters. The cluster algorithm was retained between 31.9% and 38.1% of the global dataset, corresponding to between 43.4% and 47.1% of volume transfers. Note that the clustering algorithm is only applied to stopped "samples," and that the sub-sampling and DBSCAN procedures can be tuned to different results. After association to POIs, almost all cluster samples are retained, with the exception of site A where one large cluster could not be assigned to a POI (more details on the next section).

For the samples associated to POIs, we observe a consistent behavior of throughput versus distance at most sites. Figure 4 presents per-distance boxplots of transfer rates. A fitted exponential curve $t(d) = a \cdot e^{b\,d}$, where t is throughput and d is distance, is also shown with values presented in Table 3. Sites A and C, containing samples at least up to 50 m, exhibit similar curves. Site B does not allow for extensive comparison (due to inferior distance range), but rates at the available distances verify those of other sites.

Table 2 Dataset metrics: total, stopped, and cluster-bound (produced and assigned)

Site	# days		Total	Stopped	# clus.	Clusters	POIs	Assigned clusters
A	70	# samples	4148	75.4% (3128)	21	34.6% (1434)	16.5% (685)	18
		Conn. time (s)	12,825	76.4% (9793)		31.1% (3984)	14.7% (1880)	
		Volume (Mbit)	103,006	74.4% (76,614)		45.2% (46,528)	20.0% (20,607)	
B	34	# samples	1386	81.1% (1124)	11	31.9% (442)	31.9% (442)	11
		Conn. time (s)	4094	81.6% (3340)		29.4% (1202)	29.4% (1202)	
		Volume (Mbit)	31,404	82.8% (25,995)		43.4% (13,631)	43.4% (13,631)	
C	61	# samples	6466	93.5% (6043)	10	38.1% (2466)	37.8% (2444)	8
		Conn. time (s)	19,875	93.9% (18,655)		32.4% (6444)	32.1% (6387)	
		Volume (Mbit)	170,167	90.9% (159,813)		47.1% (80,109)	46.6% (79,283)	

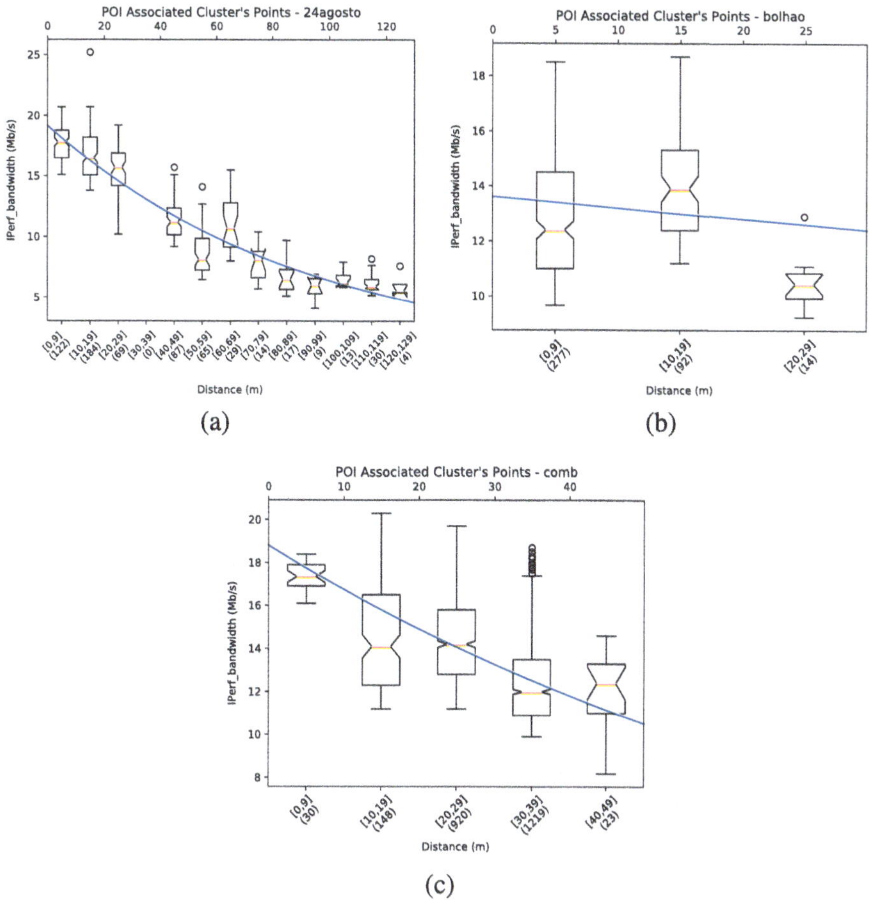

Fig. 4 Throughput vs. distance for POI-bound samples. (**a**) Site A. (**b**) Site B. (**c**) Site C

Table 3 Exponential regression of throughput vs. distance for POI-bound samples

Site	Factor (a)	Exponent (b)
A	19.19	−0.0109
B	13.64	−0.0031
C	18.84	−0.0116

4.2 Analysis per POI Class and Location

We evaluate the contribution of each POI class to the data transfers measured at each site. Table 4 indicates the number of POIs per site, the number of clusters associated to each POI class, and the corresponding dataset size and data volume. Figure 5 depicts the produced clusters and relevant POIs, to provide a visual support reference. Site-specific analyses are now drawn.

Table 4 Contribution of different POI classes for site performance

Site	POI class	# POIs	# clusters	# samples	Conn. time at class POI (s)	Volume at class POI (Mbit)
A	Traffic Lights	7	9	548	1412	17,023
	Crosswalks	7	2	60	152	1540
	Garbage bins	6	7	118	316	2044
B	Traffic lights	4	11	464	1202	13,631
	Crosswalks	1	0	0	0	0
	Garbage bins	1	0	0	0	0
C	Traffic Lights	5	7	30	377	5424
	Crosswalks	2	0	0	0	0
	Garbage bins	2	1	2310	6010	73,858

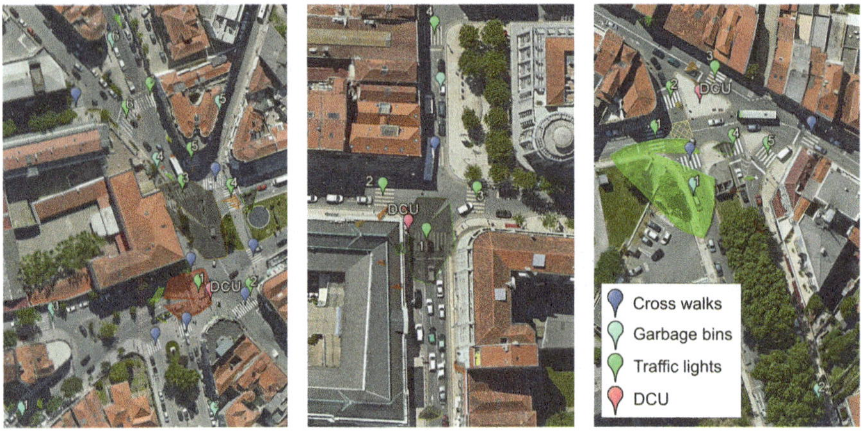

(a) (b) (c)

Fig. 5 Clusters of I2V samples and POIs (traffic lights and bins are numbered). (**a**) Site A. (**b**) Site B. (**c**) Site C

Site A: most samples and clusters are associated with traffic lights (e.g., 1, 3, 4, and 5 of Fig. 4a), and two clusters are associated with garbage bins (1 and 6). Despite being more numerous, crosswalks account for comparatively few samples. There is a large cluster that cannot be directly associated to any POI, as it is located at the center of the crossroads.

Site B: the majority of the data volume in this site is recorded at clusters associated with two traffic lights (1 and 2 in Fig. 4b). There is a garbage bin POI, but no nearby cluster had been produced during the clustering stage.

Site C: the majority of samples are associated with garbage bin 1 of Fig. 4c. Trucks stop at this bin for long periods; samples from nearby traffic light might have been incorporated.

In summary, we observe a clear relation between POIs and the areas where most throughput samples of high value are recorded. Traffic lights always show associated samples, and crosswalks account for few or no I2V samples. The identification of a cluster at the center of the crossroads, in site A, shows that other unaccounted factors may exist. Most samples at this cluster were obtained during night, thus excluding traffic jams as a cause; additional work is required to justify it. Finally, a single fleet-related POI may account of the majority of throughput samples at a given site (e.g., site C), to an extent that is not observed in traffic-related POIs.

5 Conclusions and Future Work

Using real-world I2V link measurements, we observe that most measurements with fleet vehicles occur with stopped vehicles, and areas of relevant throughput samples (many samples of high value) are close to POI that affect the vehicles' mobility. In detail, the contribution to I2V transfers of traffic lights is always present, and that of crosswalks is residual; strategically-located fleet-related POIs can contribute considerably. The end goal is to build a estimation model of I2V data volumes based on qualitative features of a potential site.

Acknowledgements This work was funded by project SmartCityMules, reference PTDC/EEI-TEL/2008/2014, project S2MovingCity, reference CMUP-ERI/TIC/0010/2014, funded by FCT/MEC (PIDDAC) through national funds, and IT internal funds, reference UID/EEA/50008/2013.

References

1. Bergamo, P., Cesana, M., Maniezzo, D., Pau, G., Yao, K., Whiteman, D., Gerla, M.: IEEE 802.11 wireless network under aggressive mobility scenario. In: Proceedings of the International Teletraffic Congress (ITC), Las Vegas (2003)
2. Enterprise, H.: White paper: smart cities and the internet of things. Technical report, HP (2016). http://h20195.www2.hpe.com/v2/getpdf.aspx/4AA6-5129ENW.pdf (last visited: 11 Jan 2017)
3. Eriksson, J., Balakrishnan, H., Madden, S.: Cabernet: vehicular content delivery using Wi-Fi. In: Proceedings of the 14th ACM International Conference on Mobile Computing and Networking, MobiCom '08, pp. 199–210. ACM, New York (2008)
4. Ester, M., Kriegel, H.P., Sander, J., Xu, X.: A density-based algorithm for discovering clusters a density-based algorithm for discovering clusters in large spatial databases with noise. In: Proceedings of the Second International Conference on Knowledge Discovery and Data Mining, KDD'96, pp. 226–231. AAAI Press, Menlo Park (1996)
5. Gass, R., Scott, J., Diot, C.: Measurements of in-motion 802.11 networking. In: 7th IEEE Workshop on Mobile Computing Systems and Applications, Aug 2006, pp. 69–74 (2006)
6. Hull, B., Bychkovsky, V., Zhang, Y., Chen, K., Goraczko, M., Miu, A.K., Shih, E., Balakrishnan, H., Madden, S.: CarTel: a distributed mobile sensor computing system. In: 4th ACM SenSys. Boulder, Nov 2006 (2006)
7. Li, M., Claypool, M., Kinicki, R.: WBest: a bandwidth estimation tool for IEEE 802.11 wireless networks. In: 33rd IEEE Conference on Local Computer Networks, Oct 2008, pp. 374–381 (2008)

8. Luis, Y., Santos, P.M., Lourenco, T., Pérez-Penichet, C., Calçada, T., Aguiar, A.: UrbanSense: an urban-scale sensing platform for the internet of things. In: 2016 IEEE International Smart Cities Conference (ISC2), Sept 2016, pp. 1–6 (2016)
9. NLANR/DAST: Iperf. https://iperf.fr/
10. Ott, J., Kutscher, D.: Drive-thru internet: IEEE 802.11b for "automobile" users. In: 23rd Conference of IEEE Computer and Communications Societies, INFOCOM 2004, Mar 2004, vol. 1, pp. 365–373 (2004)
11. Santos, P.M., Calçada, T., Sargento, S., Aguiar, A., Barros, J.: Experimental characterization of i2v Wi-Fi connections in an urban testbed. In: Proceedings of the 10th ACM MobiCom Workshop on Challenged Networks, pp. 5–8. ACM, New York (2015)
12. Santos, P.M., Calçada, T., Guimarães, D., Condeixa, T., Sargento, S., Aguiar, A., Barros, J.A.: Demo: platform for collecting data from urban sensors using vehicular networking. In: Proceedings of 21st ACM MobiCom, pp. 167–169. ACM, New York (2015)
13. Santos, P.M., Rodrigues, J.G.P., Cruz, S.B., Lourenço, T., d'Orey, P.M., Luis, Y., Rocha, C., Sousa, S., Crisóstomo, S., Queirós, C., Sargento, S., Aguiar, A., Barros, J.: PortoLivingLab: an IoT-based sensing platform for smart cities. IEEE Internet Things J. **5**(2), 523–532 (2018)
14. Wellens, M., Westphal, B., Mahonen, P.: Performance evaluation of IEEE 802.11-based WLANs in vehicular scenarios. In: 65th IEEE Vehicular Technology Conference, pp. 1167–1171 (2007)

Urban Landscape Revolution: The Potential of Connected Vehicles and Their Impact on the Mobility Ecosystem

Steven Andorka and Kira Rambow-Hoeschele

1 Introduction

The Internet of Things (IoT) is one of the most important technological developments in today's business world being defined as a new technological paradigm that aims to connect anything and anyone at any time and any place, giving rise to innovative new applications and services. Several studies estimate an enormous economic impact that IoT will create. Especially in the mobility sector, the IoT has the potential to create a substantial transformation.

The automotive industry is currently trying to find answers to four important technological trends: connectivity, autonomous driving, sharing and services, and electrification. Being one of the main drivers of this industry transformation, the IoT is of high importance not only for original equipment manufacturers but also for their suppliers as well.

Those new technological developments will directly affect a substantial part of society in the near future with 470 million connected cars driving on the roads until 2025 [1].

The conversion of vehicles to electromobility demands well-conceived concepts and strategies for new electrified mobility concepts. Not only is the impact of electromobility on power system infrastructure critical but also power system design

S. Andorka
Glasgow Caledonian University, School of Engineering, Glasgow, UK
e-mail: steven@andorka.de

K. Rambow-Hoeschele (✉)
Robert Bosch GmbH, Headquarters, Automotive Strategy, Stuttgart, Germany
e-mail: kira.rambow-hoeschele@de.bosch.com

R. José et al. (eds.), *3rd EAI International Conference on IoT in Urban Space*,
EAI/Springer Innovations in Communication and Computing,
https://doi.org/10.1007/978-3-030-28925-6_14

and operation need to be revised such that future electrified mobility concepts succeed. To ensure this, intelligent connectivity features need to be built in into the cars of the future.

In this paper, a variety of connectivity features is being discussed. However, as a large variance of features exists, this study focuses on human-centric IoT features, which are those features with which the customer has direct touchpoints and for which sensitive personal data is used. Examples are features with integrated face recognition, voice recognition, body screening, or in general features involving human interaction.

2 Technical and Economic Thoughts

2.1 General

With an increasing level of connectivity standards in cars, a variety of new features that are primarily built on personal data will be available. For some features it might be useful or even necessary to have a digital identity that incorporates private information to elaborate their full potential. However, as the usage of digital identities is a highly controversial topic, it is of importance to keep it in mind when the individual feature categories are being introduced in the following.

When discussing digital identities, it is crucial to consider that high security standards need to be ensured in order to use them without major concerns. In case of a digital identity, scientists define three ways how it could be accessed in a car: first, through what we know (passwords), second, through what we have (keycards), and third, through what we are (biometric data).

As biometric scanning, including face and voice recognition, is considered one of the safest ways to identify persons, it could be used in a variety of ways in the future as it fulfills several important criteria. First, biometric data is universal as every person possesses it. Second, it is distinctive as every human is unique. Third, it is simple to use as nothing else besides the person being present is necessary [2–4].

However, even without a biometric scanning, many features will be available in the future that will be seamlessly integrated into the car. With the help of cameras and sensors inside and outside of the car, cloud platforms, artificial intelligence (AI), smartphone integration, data from social media, and many more input variables, the car will change in a way it has never changed before [5].

To cluster the large variance of human-centric connectivity features that might be introduced in the future, five feature categories are being introduced in the following: personal assistant, health, convenience, security and safety, and legal and insurance.

2.2 Personal Assistant

The first potential transformation that the car might undertake in the future is the conversion into a highly customized personal assistant that supports its passengers in a broad variety of situations in life. The idea is that while one is using a car, one's personal assistant is constantly learning from oneself. Every individual will have its own profile which can be infused into any car one owns, rents, shares, or drives. The assistant uses context and additional smart data from different sources such as mobile devices, social media, or personal conversations and utilizes them to create recommendations to improve and facilitate the life of its users [6–9].

The assistant builds up a relationship with its user through the capability of empathic thinking leading to a more and more personal experience over time. Passengers will have conversations with their cars like they have conversations with humans today. To make this feel more natural for individuals, the assistant might even adapt its voice to, for instance, the parents' voice as people feel more comfortable around a familiar voice. Through those and other conversations, the car will be able to identify and analyze the personality of its users and will try to obtain a comprehensive picture of its passengers to serve them in the most supportive way possible [10–12].

There is a variety of application fields of the personal assistant. In the following some of the most important features are introduced. First, the integration of smart home applications is a promising field. In the future, it will be possible to actively steer functions such as lighting or heating of one's house while being in the car. Moreover, the car could either recommend going shopping, as it communicates with the refrigerator realizing that certain food is missing, or it could automatically purchase the food according to one's diet plan or meal recommendations based on one's health profile. Other exemplary features are surveillance features of one's home to grant certain people access by opening the front door or to check whether all windows are closed [13–15].

Second, the field of connected navigation and location-based services opens up a variety of possibilities to facilitate the life of car users. By combining and analyzing different input variables such as calendar entries or conversations the car might recommend optimized routes. For example, when recognizing that an important phone call is scheduled while driving, the car might select a route with the best possible phone connection.

Furthermore, various services specifically developed for electric cars will be introduced to optimize the driving experience such as intelligent charging assistants. Through community-based parking, finding a parking slot will be easier in the future as algorithms will calculate the probability of free parking spaces based on current data collected by cars at the destination, calendar entries, or other location-based information.

With the help of historic data, drivers will be able to choose routes depending on a variety of variables such as number of accidents in the past or safety ratings of regions. In addition, the car might make intelligent route proposals based on individual priorities such as picking up a coffee on one's route to work in the morning or it might suggest car sharing with other people who choose an almost identical route frequently. In addition, when localizing the car, last mile deliveries to or into the car will be made possible through the help of face scans and individual access permissions for the car [16–20].

Third, with more and more automated driving functions being available, infotainment features will gain in importance. The assistant will make personalized recommendations for entertainment such as education, music, or films. Also, one will be informed about potentially interesting events, locations, or social activities to help optimizing one's weekday or weekend schedule based on preference and personality analysis. With an extended infotainment system, passengers might receive personalized advertisements in exchange for services and will be able to purchase goods in the car which could be picked up autonomously by the vehicle when not being occupied [21–24].

2.3 Health

The second field of change are health features that will move into the car. With the help of cameras, infra-red sensors, microphones, wearables, and other sensors in the car, on one's body, or inside one's body the car will be able to measure and analyze health parameters of its passengers and to use them in various ways. In this context, it needs to be differentiated between current status monitoring functions that measure the health and the driving ability in the short-term and disease-monitoring functions that collect data over time to diagnose or predict potential diseases [25–28].

Features included in current status monitoring are the analysis of the passengers' focus on driving, their emotional status, and their physical health status. By tracking the eyelids, head pose, iris movements, mobile phone usage, or the upper body position, the car will recognize whether its driver is focused enough on driving or one should rather have a rest before continuing the ride. Then, the car might make corresponding recommendations or might suppose to switch to autonomous driving mode.

To measure the emotional status of its passengers, the car will use input variables such as driving patterns, number of occupants, tone, content, and volume of their voices, body language, facial expressions, choice and volume of music, force of car doors being closed, social media, messages, or calendar information. Based on the analyzed mood, the car will be able to adapt music, lighting, scents, temperature, and further variables to reduce stress levels of its passengers, to energize them, or to prepare them for a better sleep after arrival. Furthermore, it might give recommendations such as calling a good friend in a state of sad emotions [29–32].

To measure the current physical health status, a variety of input variables is used. Exemplary health parameters are pulse, heart rate variability, body temperature, respiration rate, apnea detection, dehydration level, blood sugar level, or blood pressure. When an emergency, such as the driver passing out or suffering an epileptic seizure, is recognized, the car can automatically switch into autonomous mode or slow down the car safely to prevent an accident. The car could then drive the passenger autonomously to the next hospital while informing the hospital of its arrival, sending important patient data in advance to save precious time.

Moreover, after an accident current patient data can be sent to ambulances and doctors prior to their arrival and in case a doctor is driving on the same route he could be notified and asked for first help. Further features include the detection of alcohol-influenced behavior or fatigue measurement to give recommendations to the driver to not start driving or to take a rest [33–41].

In the long-term, diseases can be diagnosed or predicted with the help of algorithms and individual electronic health records that are enriched at any point in time a passenger is using the car. Thus, it will be possible to make assumptions about physical as well as psychological diseases.

Through a health cloud that is supported by machine learning, medical knowledge and data can be combined to find correlations between information that might not even be known yet. Passengers could actively ask the car for advice or they could be notified only in case of a severe threat being recognized. As a result, the car could give recommendations to improve the individual health such as optimizing the daily schedule by adding sports or changing task orders to reduce stress. Also, only passive alerts could be sent to the primary care physician to keep him informed.

Among potential physical diseases that could be recognized are diabetes, irregular heart rates, hypertension, sleep apnea, atrial fibrillation, or a heart attack. Potential psychological diseases include depression, Alzheimer, or Parkinson. The empathic car could also try to find causes for diseases such as depression and talk to the passenger to improve his current state. Whereas those are only some examples of diseases that could potentially be recognized, the number of correlations of symptoms and diseases will increase with more and more data being fed into the health cloud [42–48].

2.4 Convenience

The features of the third functional area might lead to higher convenience for passengers in the car. Through biometric scanning, the car will be able to identify its passengers enabling several features. Moreover, the car will recognize the age and gender of unknown persons and will observe the number of passenger present. Passengers with a known profile could have personalized settings for the steering wheel, the mirrors, the seat position, the infotainment, the lighting, or the room and seat temperature. The latter one might also be adapted in accordance with the passengers' calendars, their clothing or the weather at their destination.

Furthermore, biometric information could serve as a digital key to authenticate the individuals accessing the car, being especially interesting for fleet management or car sharing. Remote access via an app could be given to new drivers and a haptic key would not be needed anymore. Also, when analyzing one's biometric data, the car might give recommendations to change certain settings such as the seat position to drive more convenient. Furthermore, as the driver of the car is recognized at any time, driving style profiles will be created that can be used for recommendations by the car to change specific patterns to, for instance, drive more fuel efficient [14, 35, 49–52].

2.5 Security and Safety

The fourth field of application that might be transformed due to higher connectivity standards includes security and safety features. Through cameras inside and outside of the car, people could be identified. This could lead to less theft of cars and a quicker detection of criminals using a car. With the help of geofencing and position tracking, stolen vehicles could be found, hijackings could be recognized, and parents or fleet managers might have the option to track their children's or drivers' driving behavior allowing them to oversee the car's speed, routes, driving time, or the number of passengers inside.

Moreover, any kind of financial transaction could be made safer through biometric identification inside the car. For example, this feature could be used to pay for an online purchase, to make a money transfer, or to pay for refueling the car. Furthermore, biometric identification could be used to sign contracts or other official documents digitally [11, 53–55].

2.6 Legal and Insurance

The fifth application field concerns feature linked to legal and insurance aspects. As car data could be transferred in real time at any moment, frauds could be identified and sanctioned whenever occurring. Examples for fraud detection are the recognition of over speeding, mobile phone usage, or drunk driving. As driving behaviors can be analyzed and scored considering different variables such as safety or environmental friendliness, taxation rates for cars could be adapted in accordance.

Furthermore, the analysis of driving behaviors could transform insurance models from rather static classifications into usage- or behavior-based payment models. Drivers who, for instance, drive slower and have a reduced risk of accidents need to pay lower insurance premia than drivers with riskier driving patterns. Also, due to a variety of sensors it will be easier to reconstruct accidents and to manage claims as frauds such as accidents on purpose will be revealed right away. As the car's status will be analyzed at any time and accidents or other damages will be tracked instantly, recommendations for safe routes and places could be made and insurance premia could be adapted in case of a change in usage behavior [56–60].

3 Conclusion and Future Work

To conclude, the current development of connectivity features and shared, autonomous, and electric cars is closely interlinked. Many features are beneficiary or even necessary for a successful introduction of such cars to the mass market to make them more attractive for customers. There is a large variety of connectivity features that is going to be introduced to the market. However, as many of those features require private or highly sensitive data, ethical frameworks and high cyber security standards are of importance. As this paper was primarily focused on human-centric IoT features, more research could be conducted about further connectivity features. Also, ethics and security linked to those features could be discussed.

References

1. Gartner: Gartner says the self-driving vehicle will be the first pervasive, personal robots that most consumers will experience in their lifetime. https://www.gartner.com/newsroom/id/2884317XX (2014)
2. Accenture: Biometrics and privacy: a positive match. https://www.accenture.com/_acnmedia/Accenture/Conversion-Assets/DotCom/Documents/Global/PDF/Dualpub_9/Accenture-Biometrics-Privacy-Positive-Match.pdf (2012)
3. Delac, K., Grgic, M.: A survey of biometric recognition methods. In: Proceedings of 46th International Symposium on Electronics in Marine, Elmar-2004, pp. 184–193 (2004)
4. Liu, S., Silverman, M.: A practical guide to biometric security technology. IT Prof. **3**(1), 27–32 (2001). https://doi.org/10.1109/6294.899930
5. Meola, A.: Automotive industry trends: IoT connected smart cars & vehicles. Business Insider Deutschland. https://www.businessinsider.de/internet-of-things-connected-smart-cars-2016-10?r=US&IR=T (2016). Accessed 02 May 2018
6. Cisco IBSG: Connected vehicles from building cars to selling personal travel time well-spent.https://www.cisco.com/c/dam/en/us/solutions/collateral/industry-solutions/connected-vehicles-automotive.pdf (2011)
7. Mikusz, M., Schäfer, T., Taraba, T., Jud, C.: Transforming the connected car into a business model innovation. In: 2017 IEEE 19th Conference on Business Informatics (2017). https://doi.org/10.1109/CBI.2017.64
8. Tata Motors: Tata Motors and Microsoft India collaborate to redefine the connected experience for automobile users. https://www.tatamotors.com/press/tata-motors-and-microsoft-india-collaborate-to-redefine-the-connected-experience-for-automobile-users/ (2017). Accessed 04 May 2018
9. Volkswagen AG: Volkswagen Group selects NVIDIA as key strategic partner to develop AI-Cockpit. https://www.volkswagen-media-services.com/en/detailpage/-/detail/Volkswagen-Group-selects-NVIDIA-as-key-strategic-partner-to-develop-AI-Cockpit/view/4420953/7a5bbec13158edd433c6630f5ac445da?p_p_auth=gN0zZaZN (2017). Accessed 05 May 2018
10. IBM: Audio analytics: the sounds of systems. https://www.ibm.com/blogs/research/2016/11/audio-analytics/ (2016). Accessed 04 May 2018
11. Microsoft: IoT for transportation. https://www.microsoft.com/en-us/internet-of-things/transportation (2018). Accessed 23 Mar 2018
12. Nass, C., Yen, C.: The Man Who Lied to His Laptop: What We Can Learn About Ourselves from Our Machines. Penguin Group, New York (2010)
13. Byton: Byton experience. https://www.byton.com/experience (2018). Accessed 05 May 2018

14. IAV: IAV automotive engineering. https://www.iav.com/ (2018). Accessed 10 May 2018
15. IBM: AI-connected vehicles with Watson Assistant. https://www.ibm.com/internet-of-things/industries/iot-automotive (2018). Accessed 30 Apr 2018
16. Bosch GmbH: Bosch mobility products and services. https://www.bosch.com/products-and-services/mobility/ (2018). Accessed 15 Apr 2018
17. Harman: HARMAN and Samsung are defining and driving the future of automotive experiences. https://www.harman.com/connected-car (2018). Accessed 29 Mar 2018
18. Madell, R.: Connected vehicles and IoT technology: are you ready? https://insights.samsung.com/2017/10/26/connected-vehicles-and-iot-technology-are-you-ready/ (2017). Accessed 04 Apr 2018
19. Renaudin, V., Dommes, A., Guilbot, M.: Engineering, human, and legal challenges of navigation systems for personal mobility. IEEE Trans. Intell. Transp. Syst. **18**(1), 177–191 (2016). https://doi.org/10.1109/TITS.2016.2563481
20. Toyota: Toyota CONCEPT-i. https://www.toyota.com/concept-i/ (2018). Accessed 20 Apr 2018
21. Mavadiya, M.: Your car is your wallet: connected cars and the future of Fintech. Forbes. https://www.forbes.com/sites/madhvimavadiya/2017/04/11/connected-cars-fintech/#79056cd94c35 (2017)
22. Samsung: "Integration of our secret weapon": Samsung and HARMAN executives discuss the next chapter for driving. https://news.samsung.com/global/integration-is-our-secret-weapon-samsung-and-harman-executives-discuss-the-next-chapter-for-driving (2018). Accessed 04 May 2018
23. Sony: Sony car products. https://www.sony.de/electronics/auto-entertainment-systeme (2018). Accessed 17 Apr 2018
24. Xing, Z., Parandehgheibi, M., Xiao, F., Kulkarni, N., Pouliot, C.: Content-based recommendation for podcast audio-items using natural language processing techniques. In: 2016 IEEE International Conference on Big Data (Big Data), pp. 2378–2383 (2016). https://doi.org/10.1109/BigData.2016.7840872
25. Ahmed, M., Björkman, M., Causevic, A., Fotouhi, H., Lindén, M.: An overview on the internet of things for health monitoring systems. In: 2nd EAI International Conference on IoT Technologies for HealthCare (2015)
26. Islam, S., Kwak, D., Kabir, M., Hossain, M., Kwak, K.-S.: The internet of things for health care: a comprehensive survey. IEEE Access. **3**, 678–708 (2015). https://doi.org/10.1109/ACCESS.2015.2437951
27. Philips: Accurate contactless pulse and breathing rate measurement. http://www.ip.philips.com/licensing/program/115 (2018). Accessed 17 Apr 2018
28. Saha, H., Auddy, S., Pal, S., Kumar, S., Pandey, S., Singh, R., et al.: Health monitoring using Internet of Things (IoT). In: 2017 8th Annual Industrial Automation and Electromechanical Engineering Conference (IEMECON), pp. 69–73 (2017). https://doi.org/10.1109/IEMECON.2017.8079564
29. IBM: Cars that care, chapter 2: the diagnosis – emotions lead to danger. https://www.ibm.com/blogs/internet-of-things/iot-cars-care-2-sensors/ (2017). Accessed 25 Apr 2018
30. Said, S., AlKork, S., Beyrouthy, T., Abdrabbo, M.: Wearable bio-sensors bracelet for driveras health emergency detection. In: 2017 2nd International Conference on Bio-engineering for Smart Technologies (BioSMART), pp. 1–4 (2017). https://doi.org/10.1109/BIOSMART.2017.8095335
31. Tata Elxsi: In-car health and wellness monitoring. A Tata Elxsi perspective. http://www.tataelxsi.com/Perspectives/WhitePapers/In%20car%20wellness.pdf (2017). Accessed 06 May 2018
32. Yang, J., Chen, Y., Liu, Y., Makke, O., Yeung, J., Gusikhin, O., MacNeille, P.: The effectiveness of cloud-based smart in-vehicle air quality management. In: 2016 IEEE Advanced Information Management, Communicates, Electronic and Automation Control Conference (IMCEC), pp. 325–329 (2016). https://doi.org/10.1109/IMCEC.2016.7867226

33. Connor, J., Norton, R., Ameratunga, S., Robinson, E., Civil, I., Dunn, R., et al.: Driver sleepiness and risk of serious injury to car occupants: population based case control study. Br. Med. J. **324**, 1125 (2002). https://doi.org/10.1136/bmj.324.7346.1125
34. Empatica: Empatica products for people living with epilepsy. https://www.empatica.com/index.html (2018). Accessed 03 May 2018
35. EyeSight: Automotive products. http://www.eyesight-tech.com/product/automotive/ (2018). Accessed 28 Mar 2018
36. Guo, X., Mandelis, A., Liu, Y., Chen, B., Zhou, Q., Comeau, F.: Noninvasive in-vehicle alcohol detection with wavelength-modulated differential photothermal radiometry. Biomed. Opt. Express. **5**(7), 2333–2340 (2014). https://doi.org/10.1364/BOE.5.002333
37. Lee, Y.-H., Liu, S.-T., Lin, H.-D., Tseng, W.-J.: Driver's health management system using nanosecond pulse near-field sensing technology. In: 2014 International Conference on Computer, Communications, and Control Technology (I4CT), pp. 443–446 (2014). https://doi.org/10.1109/I4CT.2014.6914223
38. Rahman, H., Ahmed, M., Begum, S.: Vision-based remote heart rate variability monitoring using camera. In: 4th EAI International Conference on IoT Technologies for HealthCare (2017)
39. Reese, H. How Tesla Autopilot drove a man with a blood clot to the hospital, and expanded the autonomous car debate. https://www.techrepublic.com/article/how-tesla-autopilot-drove-a-man-with-a-blood-clot-to-the-hospital-and-expanded-the-autonomous-car/ (2016). Accessed 06 May 2018
40. Saravanan, S., Sudhakar, P.: Telemedicine communication system in Mobile units. In: 2017 International Conference on Computer Communication and Informatics (ICCCI), pp. 1–4 (2017). https://doi.org/10.1109/ICCCI.2017.8117750
41. Strickland, E.: 3 ways ford cars could monitor your health. https://spectrum.ieee.org/the-human-os/biomedical/diagnostics/3-ways-ford-cars-could-monitor-your-health (2017). Accessed 03 Apr 2018
42. Abdelgawad, A., Yelamarthi, K., Khattab, A.: IoT-based health monitoring system for active and assisted living. In: Smart Objects and Technologies for Social Good: Second International Conference 2016, pp. 11–20 (2017). https://doi.org/10.1007/978-3-319-61949-1_2
43. Alharthi, H.: Healthcare predictive analytics: an overview with a focus on Saudi Arabia. J. Infect. Public Health. **11**, 749 (2018). https://doi.org/10.1016/j.jiph.2018.02.005
44. Barua, S., Ahmed, M., Begum, S.: Distributed multivariate physiological signal analytics for drivers' mental state monitoring. In: International Conference on IoT Technologies for HealthCare, pp. 26–33 (2018). https://doi.org/10.1007/978-3-319-76213-5_4
45. Howcroft, J., Wallace, B., Goubran, R., Marshall, S., Porter, M., Knoefel, F.. Variation in acceleration driving patterns as a measure of older adult health status. In: 2017 IEEE International Symposium on Medical Measurements and Applications (MeMeA), pp. 298–302 (2017). https://doi.org/10.1109/MeMeA.2017.7985892
46. Iskandar, A., Prihatmanto, A., Priyana, Y.. Design and implementation electronic stethoscope on smart chair for monitoring heart rate and stress levels driver. In: 2015 4th International Conference on Interactive Digital Media (ICIDM), pp. 1–6 (2015). https://doi.org/10.1109/IDM.2015.7516338
47. Kammüller, F., Augusto, J., Jones, S.: Security and privacy requirements engineering for human centric IoT systems using eFRIEND and Isabelle. In: 2017 IEEE 15th International Conference on Software Engineering Research, Management and Applications (SERA), pp. 401–406. (2017). https://doi.org/10.1109/SERA.2017.7965758
48. Singh, J., Kamra, A., Singh, H.: Prediction of heart diseases using associative classification. In: 2016 5th International Conference on Wireless Networks and Embedded Systems (WECON), pp. 1–7 (2016). https://doi.org/10.1109/WECON.2016.7993480
49. Car Connectivity Consortium: Car Connectivity Consortium achieves milestone in delivering smart phone as a car key solution. https://carconnectivity.org/press-release/car-connectivity-consortium-achieves-milestone-delivering-smart-phone-car-key-solution/ (2018). Accessed 18 May 2018

50. Jafarnejad, S., Castignani, G., Engel, T.: Towards a real-time driver identification mechanism based on driving sensing data. In: 2017 IEEE 20th International Conference on Intelligent Transportation Systems (ITSC), pp. 1–7 (2017). https://doi.org/10.1109/ITSC.2017.8317716
51. Panasonic: Automotive solutions. https://na.panasonic.com/us/automotive-solutions (2018). Accessed 15 March 2018
52. Waymo: We're building the world's most experienced driver. https://waymo.com/ (2018)
53. IBM: Automotive 2020 - clarity beyond the chaos. http://www-07.ibm.com/shared_ downloads/6/IBM_Automotive_2020_Study_Clarity_beyond_the_Chaos.pdf (2008)
54. GSMA: Automotive technology developments. https://www.gsma.com/iot/automotive/ (2018). Accessed 09 Apr 2018
55. Outay, F., Kammoun, F., Kaisser, F., Atiquzzaman, M.: Towards safer roads through cooperative hazard awareness and avoidance in connected vehicles. In: 2017 31st International Conference on Advanced Information Networking and Applications Workshops (WAINA) (2017). https://doi.org/10.1109/WAINA.2017.17
56. Harner, I.: How will IoT impact the insurance industry? https://www.iotforall.com/insurance-iot-usa-summit-recap/ (2018). Accessed 20 Mar 2018
57. Husnjak, S., Perakovic, D., Forenbacher, I., Mumdziev, M.: Telematics system in usage based motor insurance. Proc. Eng. **100**, 816–825 (2015, February 24). https://doi.org/10.1016/j.proeng.2015.01.436
58. McKinsey & Company: Connected car, automotive value chain unbound (2014)
59. Microsoft: Empowering automotive innovation - seizing the connected car opportunity with Microsoft. Retrieved from https://www.google.com/url?sa=t&rct=j&q=&esrc=s&source=web&cd=1&cad=rja&uact=8&ved=0ahUKEwj2sbjamezaAhVJtRQKHTHqCu8QFggqMAA&url=http%3A%2F%2Fdownload.microsoft.com%2Fdownload%2F6%2F9%2FD%2F69D92EB1-F1EE-4893-ABE1-C005D7F9FF57%2FMicrosoft_Connected_Vehicle (2017)
60. Vodafone: Vodafone fleet telematics. http://www.vodafone.com/business/iot/end-to-end-solutions/automotive/fleets (2018). Accessed 01 May 2018

Review of Technological and Economic Considerations on Future Vehicle Design: Autonomous, Connected, Electric, and Shared Vehicles

Steven Andorka and Kira Rambow-Hoeschele

1 Introduction

The automotive industry is currently undergoing a transformation that has the potential to change the whole industry and their business models. Connectivity, autonomous driving, sharing and services, and electrification are four important trends on which OEMs and their suppliers are currently working. While companies set different priorities on those trends, connectivity and electrification are the most pressing for many. According to several studies, 250 million connected cars will be on the road by 2020.

This number is likely to increase to 470 million until 2025 being one reason for the market volume of mobility and digital services to potentially increase to 140 billion euros. Moreover, one out of three cars could have integrated biometric sensors that support certain connectivity features by 2025 according to studies. With more and more connectivity features being present, car manufacturers aim to foster electric vehicles through offering new services that increase the attractiveness of electric cars for customers [1–5].

As more electricity will be needed with an increasing number of electric vehicles, the power system infrastructure needs to be adapted accordingly. Whereas a higher demand for electricity will challenge the infrastructure, intelligent battery management and cars communicating with their charging infrastructure and other vehicles will help to find solutions for the challenges coming with the technological

S. Andorka
Glasgow Caledonian University, School of Engineering, Glasgow, UK
e-mail: steven@andorka.de

K. Rambow-Hoeschele (✉)
Robert Bosch GmbH, Headquarters, Automotive Strategy, Stuttgart, Germany
e-mail: kira.rambow-hoeschele@de.bosch.com

R. José et al. (eds.), *3rd EAI International Conference on IoT in Urban Space*,
EAI/Springer Innovations in Communication and Computing,
https://doi.org/10.1007/978-3-030-28925-6_15

developments. This paper investigates current trends for future mobility concepts by considering economic and technical aspects critical for electrified new mobility solutions to be able to succeed.

2 Technical and Economic Considerations

2.1 Transforming Perception of the Car

With emerging connectivity features, data will be collected to further foster the development of autonomous driving. Some studies estimate that 25% of all passenger vehicles in mature markets could drive fully autonomous by 2030, providing its drivers and passengers a multitude of possibilities how they could spend time in cars. Depending on the speed of development of autonomous driving, some connectivity features will enter the market faster or slower.

With more and more autonomous functions being present, the car will transform from mainly being a mobility solution into a 3rd living space that will allow people to use their time more efficient and productive while serving them as an intelligent assistant to facilitate their lives.

One company defined this shift as a process from "building cars to selling personal travel time well-spent," implicating that "the personality of the car might even become more important than its physical shell" [6]. According to a study, 20% of car buyers would switch to a different car brand if it had better connectivity features. While this number seems comparably low, the perception of the importance of connectivity features in the car keeps increasing [3, 7–14].

2.2 Frameworks

In literature, a variety of categorizations and frameworks for connectivity features is available. Generally spoken, connectivity features will affect three areas of the car.

First, by being self-integrating and self-configuring, the interaction between the car and its occupants will change. Second, by being self-socializing and self-driving the collaboration with other vehicles and devices will transform. Third, by being self-learning and self-healing the car will take care of itself automatically [5, 11, 15].

2.3 Mobility as a Service

Instead of using the car as a simple transportation solution, the digital user experience will be of high importance for OEMs and their suppliers in the future.

By creating a seamless integration of technology into the human–machine interface, interactions with the car will feel more natural.

Mobility will become a highly personalized service with the help of enormous amounts of data that will be collected and used to create a unique and individualized experience for the customers. Whereas measures such as horsepower or acceleration were indicators of vehicle performance in the past, functionality, comfort, entertainment, computing power, or high-tech experience will be success factors of cars in the future [15–22].

2.4 *Network Requirements*

Companies forecast that cars might produce up to 4000 gigabytes of data per day when being connected and autonomous, transforming them to computers on wheels as some experts describe them. Besides handling the enormous size of collected data, further key challenges are an intelligent way of data processing and data security as much information included in those data sets is very sensitive.

Two innovative technologies that are currently developed and that will help foster car connectivity are quantum computing and the 5G network. Quantum computing will provide even faster and more efficient processing of data, while the rollout of the 5G network will ensure that Internet networks will not be overloaded, supplying more capacity as well as a higher safety and reliability standard.

Whereas nowadays, those amounts of data seem relatively large considering that they will be produced on a daily basis by millions of cars, they might not be such an issue in the future due to technological progress. According to Moore's law, computing power increases every year by roughly 60%, meaning that every 18 months, computers become more than twice as powerful. However, besides an improved performance of computers, a strong and reliable Internet network is needed as well.

A very similar law applies to the technological development of Internet bandwidth. Following Nielsen's law, the Internet bandwidth will grow by 50% every year, equivalent to doubling its speed every two years. The past has proven both laws, and science predicts the validity of both laws for the near future.

However, scientists also claim that the growth will stagnate at some point in time. Fortunately, there is a broad acceptance among scientists and companies that technology development will be fast enough, ensuring that car connectivity will not be restricted by computer power or network capabilities [23–30].

2.5 *Vehicle Communication*

Cars will be in interaction with many different objects in the future, creating a complex ecosystem that accommodates a variety of different players. This technological development is called vehicle-to-everything communication (V2X) meaning that cars will be designed to connect with nearly everything that could be of interest.

Examples of V2X communication are vehicle-to-infrastructure communication, meaning that cars will be connected to traffic lights or traffic signs; vehicle-to-vehicle communication, meaning that cars will communicate with each other to prevent accidents or drive more efficiently; or vehicle-to-network communication, meaning that cars will be connected to the cloud to receive software updates.

Moreover, even more personalized features might exist in the future such as vehicle-to-pedestrian communication, meaning that cars will communicate with pedestrians and their devices to create a more dynamic traffic flow or to pick up a person at a certain place, as well as brain-to-vehicle communication, meaning that cars will be able to recognize brainwaves and, for instance, could be steered by them [31–33].

2.6 New Application Fields

The large variety of communication channels opens up many possibilities for new connectivity features in several areas of application. Together with technologies such as augmented reality or AI, in particular machine learning, cars could be transformed into places like bureaus, gyms, entertainment centers, or doctor's offices.

Worldwide, 1.2 billion people spend approximately 50 min in their car on a daily basis. Assuming that a typical day contains 8 working hours, productivity could increase by 10–15% if this time was used more effectively. Considering that an average US citizen is stuck in traffic for 52 h/year, car connectivity features have the potential to improve the quality of time spent in the car on the one hand and to decrease the time being stuck through better traffic management on the other hand [31, 34, 35].

One interesting application field of car connectivity features is passenger health. According to a study, investments into healthcare IoT solutions will reach up to one trillion dollars by 2025, improving the access to highly personalized and on-time healthcare services for people. This might also open up possibilities for health solutions in the car through new technological developments.

Already today, it is possible to analyze the mood of the passengers, which can be especially helpful for accident prevention as studies show that the likelihood of an accident increases by a factor of ten when showing emotions. After recognizing the situation, the car could start countermeasures to improve the mood of the driver and thus would decrease the probability of an accident.

With autonomous driving and health connectivity features, the car will have the ability to prevent accidents and to save lives by, for instance, switching into autonomous mode and driving to the next hospital when recognizing a driver's steering incapacity [36–39].

3 Conclusion and Future Work

To conclude, the car is about to go through a transformational process in a variety of areas. While this transformation will provide many new possibilities regarding the usage of cars, several challenges such as cybersecurity or network capacity need to be worked on. This thesis aimed to provide insights on technical and economic aspects that need to be considered when new technologies in the fields of autonomous, connected, electric, and shared vehicles are being developed in order to ensure that new technologies in cars are able to succeed. This paper should give an overview on topics that are important to consider. However, this paper is not exhaustive. As this is a constantly developing technology field, many questions are still unanswered. One interesting further research topic could be, for instance, deeper research on cybersecurity challenges that are linked to technological progress in the car.

References

1. Capgemini: AutomotiveConnect: driving digital. https://www.capgemini.com/consulting/wp-content/uploads/sites/30/2017/07/automotiveconnect_driving_digital_pov_2016-07-26.pdf (2016)
2. Frost & Sullivan: Biometrics in the global automotive industry, 2016–2025 (2016)
3. Gartner: Gartner says the self-driving vehicle will be the first pervasive, personal robots that most consumers will experience in their lifetime. https://www.gartner.com/newsroom/id/2884317 (2014)
4. Gartner.: Gartner says by 2020, a quarter billion connected vehicles will enable new in-vehicle services and automated driving capabilities. https://www.gartner.com/newsroom/id/2970017 (2015)
5. Strategy&.: In the fast lane - the bright future of connected cars. https://www.strategyand.pwc.com/media/file/Strategyand_In-the-Fast-Lane.pdf (2014)
6. Cisco: Connected vehicles - from building cars to selling personal travel time well-spent. https://www.cisco.com/c/dam/en/us/solutions/collateral/industry-solutions/connected-vehicles-automotive.pdf (2011)
7. Little, A.D.: The future of automotive mobility. http://www.adlittle.de/sites/default/files/press-release/adl_future_of_automotive_mobility_short.compressed.pdf (2017)
8. Boston Consulting Group: Building the digital car company of the future. http://img-stg.bcg.com/BCG-Building-the-Digital-Car-Company-of-the-Future-Apr-2017_tcm9-151175.pdf (2017)
9. Deloitte: The future of mobility: what's next? https://www2.deloitte.com/content/dam/insights/us/articles/3367_Future-of-mobility-whats-next/DUP_Future-of-mobility-whats-next.pdf (2016)
10. Hyundai.: How will the Internet of Things transform the car industry? https://www.hyundai.news/eu/technology/how-will-the-internet-of-things-transform-the-car-industry/ (2017). Accessed 04 May 2018
11. McKinsey & Company: Connected car, automotive value chain unbound (2014)
12. McKinsey & Company: How the convergence of automotive and tech will create a new ecosystem. https://www.mckinsey.com/industries/automotive-and-assembly/our-insights/how-the-convergence-of-automotive-and-tech-will-create-a-new-ecosystem (2016). Accessed 15 April 2018

13. Wyman, O.: Automakers in a digital world. http://www.oliverwyman.de/content/dam/oliver-wyman/v2/publications/2017/jan/20170119_Digital_OEM_Automakers_in_a_Digital_World_web_final.pdf (2017)
14. Yogarayan, S., Azman, A., Razak, S., Raman, K., Abdullah, M., Ibrahim, S., et al.: A study of cloud based connected car services. In: 5th International Conference on Information and Communication Technology (ICoIC7), pp. 1–6 (2017). https://doi.org/10.1109/ICoICT.2017.8074645
15. IBM: Automotive 2025: industry without borders. https://www-935.ibm.com/services/multimedia/GBE03640USEN.pdf (2015)
16. Hossai, M., Shahjalal, M., Nuri, N.: Design of an IoT based autonomous vehicle with the aid of computer vision. In: International Conference on Electrical, Computer and Communication Engineering (ECCE) (2017). https://doi.org/10.1109/ECACE.2017.7913003
17. IBM: Advancing mobility - the new frontier of smarter transportation. https://www-01.ibm.com/common/ssi/cgi-bin/ssialias?infotype=PM&subtype=XB&appname=GBSE_GB_TI_USEN&htmlfid=GBE03375USEN&attachment=GBE03375USEN.PDF (2010)
18. Intel: The Internet of Things (IoT) and automotive & transportation policy principles. https://www.intel.com/content/dam/www/public/us/en/documents/corporate-information/policy-iot-automotive-transportation.pdf (2014)
19. Microsoft: Microsoft connected vehicle platform helps automakers transform cars. https://blogs.microsoft.com/blog/2017/01/05/microsoft-connected-vehicle-platform-helps-automakers-transform-cars/ (2017). Accessed 04 May 2018
20. Microsoft.: Empowering automotive innovation - seizing the connected car opportunity with Microsoft. https://www.google.com/url?sa=t&rct=j&q=&esrc=s&source=web&cd=1&cad=rja&uact=8&ved=0ahUKEwj2sbjamezaAhVJtRQKHTHqCu8QFggqMAA&url=http%3A%2F%2Fdownload.microsoft.com%2Fdownload%2F6%2F9%2FD%2F69D92EB1-F1EE-4893-ABE1-C005D7F9FF57%2FMicrosoft_Connected_Vehicle (2017)
21. Mikusz, M., Schäfer, T., Taraba, T., Jud, C.: Transforming the connected car into a business model innovation. 2017 IEEE 19th Conference on Business Informatics (2017). https://doi.org/10.1109/CBI.2017.64
22. Samsung: Integration of our secret weapon: Samsung and HARMAN executives discuss the next chapter for driving. https://news.samsung.com/global/integration-is-our-secret-weapon-samsung-and-harman-executives-discuss-the-next-chapter-for-driving (2018). Accessed 04 May 2018
23. Amon, C., Lu, D., Vestberg, H.: Mobile innovation: how 5G will enable the future. (S. Marek, Interviewer) (2018)
24. Intel: Data is the new oil in the future of automated driving: https://newsroom.intel.com/ editorials/krzanich-the-future-of-automated-driving/?_ga=2.180387495.1615764630.1522750481-587785183.1522487548 (2016). Accessed 04 May 2018
25. Moore, G.: Cramming more components onto integrated circuits. Electronics. **38**(8), 114 (1965). https://doi.org/10.1109/N-SSC.2006.4785860
26. Moore, G.: Lithography and the future of Moore's Law. IEEE Solid State Circuits Soc. Newsl. **11**(3), 37–42 (2006). https://doi.org/10.1109/N-SSC.2006.4785861
27. Moore, G.: Progress in digital integrated electronics. IEEE Solid State Circuits Soc. Newsl. **11**(3), 36–37 (2006). https://doi.org/10.1109/N-SSC.2006.4804410
28. Nielsen, J.: Nielsen's law of internet bandwidth. https://www.nngroup.com/articles/law-of-bandwidth/ (1998). Accessed 05 May 2018
29. Strandberg, K., Olovsson, T., Jonsson, E.: Securing the connected car: a security-enhancement methodology. https://ieeexplore.ieee.org/xpl/RecentIssue.jsp?punumber=10209, vol. 1, pp. 56–65 (2018). https://doi.org/10.1109/MVT.2017.2758179
30. Uppoor, S., Trullols-Cruces, O., Fiore, M., Barcelo-Ordinas, J.: Generation and analysis of a large-scale urban vehicular mobility dataset. IEEE Trans. Mob. Comput. **13**(5), 1061–1075 (2013). https://doi.org/10.1109/TMC.2013.27

31. McKinsey & Company: Car data: paving the way to value-creating mobility - perspectives on a new automotive business model. https://www.mckinsey.de/files/mckinsey_car_data_march_2016.pdf (2017)
32. Outay, F., Kammoun, F., Kaisser, F., Atiquzzaman, M.: Towards safer roads through cooperative hazard awareness and avoidance in connected vehicles. In: 2017 31st International Conference on Advanced Information Networking and Applications Workshops (WAINA) (2017). https://doi.org/10.1109/WAINA.2017.17
33. Pressas, A., Sheng, Z., Fussey, P., Lund, D.: Connected vehicles in smart cities: interworking from inside vehicles to outside. In: 2016 13th Annual IEEE International Conference on Sensing, Communication, and Networking (SECON) (2016). https://doi.org/10.1109/SAHCN.2016.7732976
34. IBM: A new relationship - people and cars. https://www-935.ibm.com/services/multimedia/A-new-relationship-Exec-Report-v3.pdf (2016). Accessed 06 May 2018
35. Tata Elxsi: In-car health and wellness monitoring. A Tata Elxsi perspective. http://www.tataelxsi.com/Perspectives/WhitePapers/In%20car%20wellness.pdf (2017). Accessed 06 May 2018
36. Chang, D., Chiu, D., Ramachandran, M., Li, C.-S.: Internet of things, big data and complex information systems: challenges, solutions and outputs from IoTBD 2016, COMPLEXIS 2016 and CLOSER 2016 selected papers and CLOSER 2015 keynote. Futur. Gener. Comput. Syst. **79**, 973–974 (2018). https://doi.org/10.1016/j.future.2017.09.013
37. Knapton, S.: Which emotion raises the risk of a car crash by nearly 10 times? The Telegraph. https://www.telegraph.co.uk/news/science/science-news/12168472/Which-emotion-raises-the-risk-of-a-car-crash-by-nearly-10-times.html (2016). Accessed 06 May 2018
38. Pat, B.: How automakers can remain competitive in the era of Uber. https://www.forbes.com/sites/sap/2017/02/20/how-automakers-can-remain-competitive-in-the-era-of-uber/#2f2f1c7e7e34 (2017). Accessed 06 May 2018
39. Reese, H.: from How Tesla Autopilot drove a man with a blood clot to the hospital, and expanded the autonomous car debate. https://www.techrepublic.com/article/how-tesla-autopilot-drove-a-man-with-a-blood-clot-to-the-hospital-and-expanded-the-autonomous-ca (2016). Accessed 06 May 2018

Index

R. José et al. (eds.), *3rd EAI International Conference on IoT in Urban Space*,
EAI/Springer Innovations in Communication and Computing,
https://doi.org/10.1007/978-3-030-28925-6

The manufacturer's authorised representative in the EU is Springer Nature Customer Service Centre GmbH, Europaplatz 3, 69115 Heidelberg, Germany. If you have any concerns regarding our products, please contact ProductSafety@springernature.com

Printed and bound by CPI Group (UK) Ltd, Croydon, CR0 4YY
30/09/2026
02225239-0001